LeRoy Cook

TEACHING FLIGHT

Guidance for Instructors Creating Pilots

AVIATION SUPPLIES & ACADEMICS
NEWCASTLE, WASHINGTON

Teaching Flight
by LeRoy Cook

Aviation Supplies & Academics, Inc.
7005 132nd Place SE
Newcastle, Washington 98059-3153
asa@asa2fly.com | asa2fly.com

Visit the ASA website at *www.asa2fly.com/reader/tchflt* for the "Reader Resources" page containing additional information and updates relating to this book.

© 2019 Aviation Supplies & Academics, Inc.

All Rights Reserved. No part of this publication may be reproduced, stored in a retrieval system, or transmitted in any form or by any means, electronic, mechanical, photocopy, recording, or otherwise, without the prior written permission of the copyright holder. While every precaution has been taken in the preparation of this book, the publisher and LeRoy Cook assume no responsibility for damages resulting from the use of the information contained herein.

   None of the material in this book supersedes any operational documents or procedures issued by the Federal Aviation Administration, aircraft and avionics manufacturers, flight schools, or the operators of aircraft.

ASA-TCHFLT
ISBN 978-1-61954-849-7

Printed in the United States of America
2023 2022 2021 2020 2019   9 8 7 6 5 4 3 2 1

Cover photos: LeRoy Cook
All photographs provided by the author and used with permission.

Library of Congress Cataloging in Publication Control Number: 2019006899

CONTENTS

	BIO	iv
	FOREWORD	v
	INTRODUCTION	vii
CHAPTER 1	FUNDAMENTAL FOUNDATION	1
CHAPTER 2	CHANGING PACE	11
CHAPTER 3	TRACING A PATH	23
CHAPTER 4	GAUGES AND GADGETS	31
CHAPTER 5	CIRCUITS AND NON-BUMPS	39
CHAPTER 6	THE ART OF ARRIVAL	47
CHAPTER 7	ALONE AND ASSURED	57
CHAPTER 8	PUSHING ON	65
CHAPTER 9	THE WORLD BEYOND	75
CHAPTER 10	THE PROVING RUN	85
CHAPTER 11	BABY STEPS, BIGGER STEPS	93
CHAPTER 12	IN PURSUIT OF KNOWLEDGE	101
CHAPTER 13	ESCAPING WEATHER	109
CHAPTER 14	DEMONIC DARKNESS	117
CHAPTER 15	EMERGENCIES	125
CHAPTER 16	IN PURSUIT OF PERFECTION	133
CHAPTER 17	THE FINAL TEST	141
	INDEX	147

BIO

A lifelong student of aviation, LeRoy Cook is an experienced pilot and instructor who's been flying and teaching for more than 50 years. He holds ATP certification for single and multi-engine airplanes and commercial certification for gliders and seaplanes. His Gold Seal flight instructor's certificate has ratings for single-engine and multi-engine airplanes, instrument (airplane), and glider. Cook is the author of over 1,700 magazine articles and has written or co-authored four aviation books, including *Beyond Flight Training*, *Flying the Light Retractables*, and *Caravan: Cessna's Swiss Army Knife with Wings*. He wrote the "CFI" column in *Private Pilot* magazine for 34 years and was the editor of *Twin & Turbine* magazine from 2012 to 2016.

FOREWORD

As a young boy, LeRoy Cook wandered onto an airfield with a dirt strip and taildragger airplanes. General aviation was blessed. LeRoy learned to fly. He got his private certificate, commercial, flight instructor, instrument and ATP…as have so many others.

But, LeRoy was different.

He never stopped learning. LeRoy Cook, after 60 years of flying, plus or minus a year or two, is still learning. He is still exploring the magic of lift, the symmetry of balanced flight, the mystery of the perfect landing. It is a personal quest. But, he is happy to share that quest with those that share his love of flight.

But that is only part of the blessing. LeRoy Cook has written numerous articles and books about flying. They range from the techniques of piloting an airplane to the joys of being in the air. His writing is infused with a quiet, plain-spoken philosophy that encourages flyers to do their best in whatever it is that they fly.

LeRoy's style of writing is lean. He never uses two words when one will do. The result is a short sentence packed with information. So, readers, pay attention. In this, his latest book, LeRoy is writing to new flight instructors. But it is must reading for any flight instructor or anyone that might aspire to become one. As a flight instructor and a Designated Pilot Examiner, I attest to the fact that the information in this book is sorely needed in the aviation community.

In this, the age of glass cockpits and the miracle of the magenta line, we are producing electronic data managers. As a result, stick and rudder skills have deteriorated. Our leading cause of accidents is classified as "Loss of Control." Many of today's flight instructors were trained to be electronic data managers. It is no surprise that that is what they teach. Keeping the ball in the center is secondary or even tertiary to the colorful screens in the cockpit. Traffic advisories send them heads down in the cockpit searching for possible airborne conflicts, when their eyes and attention should be on the horizon.

LeRoy Cook does not disparage the advanced avionics that are part of aviation today. But he does emphasize the basics of flight and the eclectic craft of teaching those basics.

Read LeRoy Cook's book. If you are a flight instructor or aspire to become one, this book will be a trusted guide.

David Bradley
CFI and DPE

INTRODUCTION

AN INTRODUCTION TO INSTRUCTING

A few years ago, when I passed the mark of 10,000 hours of dual given, it was suggested that I write a book about learning to fly. I hesitated, because there have been many, many books written about flight training. What could I possibly add that has not been said before? However, I have had the advantage of a long perspective, passing from the age of dirt runways, tailwheel trainers, tube-type radios and minimal instrumentation, to the development of today's full range of advanced aircraft. Even in this day of cockpit displays and composite airframes, the sky does not change, only our means of participating in flight.

In the course of more than 50 years of flight instructing, I have taught only a few hundred people to fly. One might expect the total to run into thousands, but the reality of aviation is that we have to introduce numerous people to flying in order to make a single licensed pilot. I once analyzed the student record folders accumulated from a dozen years of instructing; of those sticking with the program long enough to make their first solo flight, only a third continued on to gain their private license.

What happened to the other two-out-of-three? The reasons, or I should say the excuses, varied widely, but most of them boiled down to loss of interest. In some cases, a relocation interrupted the student's flying lessons; there's always the hope that those who left the local area eventually completed their training elsewhere. The fact remains, learning to fly is more work than some people want to deal with. Most who shared their feelings used the excuses of spousal impediment, poverty, lack of time and just plain fearfulness. It takes considerable effort, both in the air and with ground study, to complete the course, and we must not assume it's for everybody.

Flying, unlike many other activities, cannot be evaluated from afar. There has to be an initial period of participation before a decision can be made about continuing to devote time and treasure to the training.

Thus, a large dropout percentage is normal. Dilettantes need not apply. I try to explain the commitment required during the initial Q&A session, but there still has to be a few experimental hours flown to test the waters. I can't always judge who will last and who will wash out. I like students to say things like, "this is great!" and, "I don't want to stop!" But when the real work of building layers of piloting skill is undertaken, I know there will be hours that test resolve.

And so, we instructors have to be innovative to keep as many students as possible hungering for more. I can't teach them all, nor should I expect to. Personality conflicts aside, the sky is not for everyone, the responsibilities of a pilot-in-command are demanding, and if a person doesn't want to be there, they should not be coerced to continue. I don't often call up truant students to beg them to make an appointment. If they want to continue their lessons, they know where to find us. That said, everyone needs encouragement at various times, and we CFIs must not lose sight of our role in encouraging and facilitating.

As a part-time CFI in a limited market, I only give a few hundred hours of instruction each year, and much of that is recurrent training and advanced schooling. But, I take my greatest satisfaction from primary flight instruction, when I can mold raw clay into the image of an aviator, give it a tinge of the passion I feel, and know that I was responsible for everything that new pilot knows about flying. I can't command that same measure of parental pride when helping a pilot gain a commercial rating or make a transition to instrument flying. Overseeing a pilot's induction into the solemn priesthood of CFI certification is close, but that's more of a passing of the torch, rather than attendance at a birth.

LESSON PLANS

The employment of formally-constructed lesson plans varies considerably from instructor to instructor, and from student to student. Innovation is the hallmark of an effective teacher, and strict adherence to a written plan of instruction limits such creativity. Nevertheless, one certainly has to have an organized plan of action when setting out to teach; something to refer to, as a means of making sure everything was covered. Just writing down the order of the steps, reading it over, and reviewing the plan after the flight, may be enough. After some years in the right seat, lesson plans become pretty well ingrained.

Most importantly, the instructor has to stay flexible. The cramped, noisy, jostling cockpit is a lousy classroom, and a very expensive one. We cannot waste time pursuing rigid monkey-motion that may not be applicable to this student on this day. If they aren't grasping the concept being introduced, drop back to a simpler previously-used maneuver that has been mastered, then work up to the advanced steps. If the day turns unsuitable, because the ceiling drops or the wind turns sideways, switch lesson plans to cover something of equal value.

In truth, lesson plans don't have to be complicated. At the start, put down the objectives of the lesson, then state the means by which these objectives will be taught, and then show how we are to know the objective has been met; can the student perform within tolerances, can they explain the maneuver, can the task be performed unaided? The technique used can vary, but the objectives remain the same. Remember, lesson planning doesn't have to be complicated.

CURRICULUM

Teaching flying requires a steady addition of more challenging material after simple, basic maneuvers are learned. Think of it as building a brick wall; lay down the first course of bricks, tamped carefully in place and aligned correctly, then follow with additional bricks on top of that foundation. You must not advance to complex maneuvers until the fundamentals are mastered.

To make the process of attaining a pilot license less daunting, I speak about doing it in phases. The first phase is to achieve solo flight status, which includes spending time practicing maneuvers without the instructor. Phase two is the cross-country section, when dual cross-countries are followed by solo trips, developing confidence and meeting all the requirements for the license. And phase three is the finishing-up segment, preparing for the flight test, a.k.a. "the checkride," by polishing and perfecting everything it takes to be a successful certificated pilot. Preparing for and passing the Knowledge Exam, a.k.a. "the written," is a phase of its own, inserted whenever it's appropriate. If there is going to be a need to halt flight training temporarily or switch instructors, it's best to do it after completing one of the phases, to minimize costly refresher training.

PACE YOURSELF

How many students can you handle? How many hours should you instruct per day? Only as many as you can enjoy. I usually supervise six, sometimes a dozen, active students—ones that fly regularly—and I try to have them spread out in various phases; two in pre-solo, a couple on supervised solo, two more in cross-country and one or two in the checkride-prep stage. Such variation helps maintain one's sanity and focus. Trying to fly 1,000 hours of dual per year guarantees burn-out. Stick with a limited number of students to make it fun. But do not put your CFI certificate into hibernation once you have a career established; keep instructing to pass along your expertise, and you will learn from ever-inquisitive students, who will challenge you to keep growing.

CONDUCT

As a certified flight instructor, conduct yourself in a professional manner. Do not demonstrate risk-taking, lest lesser-experienced eyes be watching your example. Hold to higher standards of safety, suggest better alternatives and never let your frustration with inept students show in word or action. Always empathize, remembering how you felt when you were a student pilot.

Pay attention to grooming and hygiene, given the close confines of your classroom. Body odor, perspiration and expectoration are unavoidable byproducts of cockpit activity, but must be minimized. Dress one level above your student, to set the right tone of authority: If they wear blue jeans, wear dress pants; their polo shirt requires that you wear a dress shirt; if they show up in a dress shirt, you add a tie; if the student wears a tie, you should wear a blazer. Dress modestly, never to draw attention, but in business-like attire.

Even in the intimacy of an airplane cockpit, you must respect the student's space. Watch for discomfort and avoid acts that can be misunderstood, particularly with mixed-gender or age-gap situations. While it's normal to become friends with students, particularly those with similar non-aviation interests, NEVER indulge in dating or close social activities. This only interferes with the learning process. If truly interested in pursuing a cute student, wait until the license has been attained, or hand them off to another CFI.

INNOVATE, INNOVATE

The purpose of this book is to set down my methods of teaching primary flight instruction, acquired from dozens of instructors I have known, and mingled with my own experience. Bear in mind, no one technique works for every student, the CFI's job is to evaluate the effectiveness of their own teaching and to keep trying new things to help the student overcome an obstacle to learning. Dogma is a luxury applicable to simpler pursuits; innovation is key to the transfer of knowledge.

Never demand that every student become a clone of your assessment of what it is to be a pilot. Each individual must eventually seek their own path, whether it's flying for fun, for business or personal travel, for growth or release, or to have a fulfilling career. Most graduated students will be satisfied to remain at a level you find incomplete, yet it will be enough for them. I have trained only a few totally dedicated disciples. If you are lucky enough to receive one of those for a student, enjoy the experience.

In this spirit, I will simply discuss what this half-century of pedagogical airmanship has taught me, particularly as it relates to conducting a course in flight training, and I'll leave it to you to judge if it was worthwhile. For me, it most certainly has been.

CHAPTER 1

FUNDAMENTAL FOUNDATION

As said in the introduction, any structure must be built up on a solid foundation. Each row of bricks must be firmly tamped in place, settled and evenly aligned, before the next layer is added. Thus it is with learning a skill like flying; the fundamentals are the foundation upon which all else rests. We cannot proceed on to advanced work until we're familiar with the basics.

In the first hour or two of flight instruction, we'll set the tone for an entire career of flying. In addition to the fundamentals of flight, however, it's necessary to spend time explaining the cryptic confusion of cockpit management, what the instruments are telling us and how they are used. This training can be started on the ground, but a lot of it only makes sense when it's demonstrated in the air. Let's cover the fundamentals first, then round out the introduction with supplemental subject matter.

THE FOUR FUNDAMENTALS OF FLIGHT

No, the four fundamentals are not to be described as stall, spin, crash and burn. Gallows humor joking is a time-honored pastime in aviation, but we must avoid such pointless confusion here. Students are often laboring under some apprehension already, thus we must teach them how to manage risk to enhance safety, not dwell on poor outcomes.

In truth, there are only four things you can do with an airplane: climb, glide (descend), turn and fly straight and level. All else is made up of these four fundamental maneuvers, perhaps combined with one another or chained together, but they must be learned so well that they come automatically, as with driving a car while talking to a passenger. You don't think about turning the steering wheel to round a corner, it's just an automatic response to following the curve of the road. And so it will be in flying, with practice.

The advantage one has when learning to drive, however, is that we saw our parents move the wheel, shift gears, brake and accelerate, from the time we started riding with them. It lost its mystery long before it was our turn to try it. Flying, however, is seldom learned in this way. We come to it entirely unschooled, and students have to adjust to its strange language, uncomfortable feelings, strident sounds, and unfamiliar gauges and controls.

The order of demonstrating things as they occur usually introduces a climb as the first fundamental maneuver. As we climb away after takeoff, the nose-high attitude of the aircraft is pointed out, and the best rate of climb airspeed is shown on the dominant instrument. Then, the student can see that raising the nose, by pulling back on the control wheel (or stick, or yoke, or whatever you call it), produces a response, with a slight delay, of a slower airspeed. Lowering the nose attitude gives the opposite response; airspeed increases after the nose goes down. Only one attitude is correct, the one that results in the best climb performance.

By now, there's frequently a need to turn the aircraft away from the runway heading to depart the traffic pattern. This generates the next teaching moment, illustrating how the turn is initiated with some aileron input and how the relatively shallow bank angle is stabilized by neutralizing the control. The turn continues as long as the bank is maintained, just as a bicycle leans when rounding a corner. When the desired direction is reached, opposite aileron lifts the lowered wing back to wings-level and the turn stops. Simple, no?

But, how do we know the wings are level? Look out at the wingtips, one after the other. There should be equal distance between wingtip and horizon on each side. No bank, no turn. At this point, I like to illustrate the effect of P-factor in the climb, showing that a slow progression into a left turn occurs when all control pressures are released (assuming a right-rotation tractor propeller). Then, I show that the merest pressure on the right rudder pedal stops this left-turning tendency, which is strongest at low airspeed and essentially disappears in level flight.

As we reach a safe maneuvering altitude, the nose is lowered to stop the climb, demonstrating the change in nose attitude relative to the horizon, and power is reduced to a cruise setting as airspeed builds and emphasis is shifted to the altimeter, which now becomes our primary performance instrument. Ah, but we haven't retrimmed the aircraft. I release the cleverly-concealed forward pressure I'm holding on the yoke and the nose starts upward, as the airplane attempts to seek its trimmed speed, which was back at best-climb. I then demonstrate that adding some nose-down trim allows us to achieve hands-off flight, as we did in the climb.

This is the proper time to demonstrate the $E=MC^2$ of aviation. This begins with the concept of attitude (the relationship between the aircraft's nose, as seen from the pilot's seat, and the natural horizon line, where the earth meets the sky) and adds the variation of power to achieve performance. To be forever inscribed in the student's cerebrum, we chant "Power Plus Attitude Equals Performance" as we write it across the cockpit on a chalkboard of air. Pull the nose up, but with the throttle reduced, and no climb results; it's plain to see, half of the equation is in error. Add full throttle, but with the nose on the horizon, and only noise happens—there is no climb. Again, one half of the required inputs is wrong. Only when BOTH power and attitude are correct does the desired performance take place, something that must be learned for each maneuver we use in our flying.

To complete the four fundamentals, we then demonstrate the glide maneuver, as an antithesis to the previously-learned climb. What goes up eventually has to come down, and there's a proper, precise way to descend for a landing, just as there's a way to maximize climb performance. The power setting is reduced to dead idle, to emphasize that the engine is not totally necessary for flight. We note that the nose is now heavily weighted toward dropping over into a hands-off dive, so we have to oppose this nose-heaviness with back pressure on the yoke until a handful of trim adjustments are made. By now, the airspeed has slowed to the best-glide number and we point out the nose attitude, which is below level-flight position. Obviously, we're traveling down a hill, but at a stable rate of descent, in full control. There is, then, no immediate danger when the engine's power is taken away, although an eventual landing will result.

The point made, power is reapplied and trim reset to level cruising flight. The student has now flown in all four regimes of flight; climb, turn, straight-and-level, and glide. That's all there is to know, we stress; all else is based on those four basic maneuvers. Not only can you do it, it's been proven; you've already done it. All you need now is practice and development.

Inquiring minds often want to know, before the first flight, "how much of the flying will I get to do?" My answer is always "most of it," because we're not out for a ride, but rather for a flying lesson. There should be no wasted minutes, so from the time the checklist is picked up, the student should be involved in a tactile sense. The first time out, I'll hold the checklist and read it with the student, but they will move the control or switch in response. Once we're safely underway and I've demonstrated the proper ground steering technique, taxi control is passed to the student and they learn how to follow a yellow line, slow down for turns, apply the brakes and watch out for the protruding wingtips.

Instruction in ground handling continues throughout the pre-solo phase and beyond. Often given short shrift in training, control of the aircraft on the ground should be honed continually. While it's easy to learn how to steer a tricycle gear airplane, sloppy taxiing creeps in unless the instructor insists on correct technique. Even when they're inert, one must hold onto the flight controls, in anticipation of encountering a wind gust or prop wash. Taxi on the centerline, not in a parallel universe off to one side, and reduce speed for turns by throttling back in advance to minimize the need for brakes. Brakes, I always maintain, are a substitute for brains; plan ahead so you won't need them (except for airplanes built without nosegear steering).

Concepts for ground handling are foreign to nearly all beginning students; the throttle does not snap closed on its own, like a car. Rather, it requires a manual pull to achieve an idle. The primary steering control on the ground is accomplished with one's feet, with the hand-operated controls relegated to a secondary role. In the air, the situation is reversed—hands are primary, feet are secondary, and during takeoff and landing, there is a shift from one system to the other. A beginning student has to absorb all this, and they need to be assured that it's normal to spin the control yoke frantically from one side to the other until foot-steering becomes second nature, which can take five lessons or more.

The first takeoff or two are a shared responsibility; the student has already learned how to taxi, so following a runway centerline while the airplane accelerates toward liftoff is no mystery. I will take care of raising the nose when airspeed is sufficient, and point out that the airplane lifts off entirely on its own, the result of speed creating lift with the wing at a high angle of attack. By the time we've accelerated to V_Y, where the airplane will be in trim, I can turn the controls over to the student and begin the teaching of fundamentals. The next time we take off, I'll track the centerline for the student, while they raise the nose and fly through the liftoff, and if all goes well the third takeoff is all theirs.

This does not mean, of course, that instructors can retreat from responsibility, most particularly when flying close to the ground. Never, ever, doze with unguarded controls when the runway is near. Even students with several lessons under their belt can react unexpectedly, and it only takes an instant to damage a nosewheel, depart the runway edge or zoom upward into a stall. I maintain, only half facetiously, that students teach themselves to fly. I am only there to protect life, limb and property, and to shorten the process by explanation, demonstration and tedious repetition.

CLEAN-UP WORK

The first few lessons are where we must slip in the tidbits of instruction that have to be learned to foster understanding. Once the process of turning is learned, by entering a bank and rolling out of it, we have to answer the unvoiced question, "but what about the rudder? I thought it had something to do with turning the airplane." And so it does, but only in a secondary role, I say. Its primary purpose is to swing the nose from side to side, something that's not normally desired. However, a turn begun without rudder input, I demonstrate, is a sloppy, hesitant maneuver, and the aircraft occupants are tossed sideways until the turn stabilizes. A similar roughness takes place during rollout. The rudder, then, must be used to overcome the airplane's natural tendency to keep going where it's pointed, if the turn is to be made smoothly. The inclinometer is evidence of this characteristic. When I show a properly coordinated turn entry, with the inclinometer ball centered, the turn begins immediately, no discomfort is felt by the passenger seated in row 27, and the subsequent rollout is exactly on the desired heading, with rudder assisting aileron to stop the turn. Now, to make them understand that one's legs are much more powerful than one's arms, and only a tiny rudder input is needed...

Because we rely so much on the directional gyro for heading information, it's important to show the reason we have it installed in the middle of the panel, with the master "wet" compass far off, out of view. The magnetic compass is a reliable source of direction, with no power to fail, but it's reliable only in level, steady flight; the student soon sees that it can't be trusted in a turn, and that it wobbles in rough air. The directional gyro, on the other hand, needs periodic resetting, particularly after aggressive maneuvering, but it does couple to the aircraft's heading during a turn.

I always re-emphasize the importance of attitude control during turns, by asking the student to hold a concealing checklist to block off only my view of the altimeter while I perform a 180-degree turn. He or she sees that I'm able to manage a turn without varying altitude more than a few feet, so obviously I'm not using the altimeter's needle to make a steady turn. In doing so, I show the left and right level-flight nose attitudes that should be our primary reference, with the altimeter used only to grade performance.

And so goes the introduction to the fundamentals of flying, encompassing the four basic maneuvers used throughout aviation, as well as the basic flight instruments, power settings and ground handling. At this point, the student feels they are really and truly flying an airplane, and that this might not be so mysterious after all. Ah, but it's only the beginning...

LESSON PLAN

OBJECTIVE

An introduction to the fundamentals of flight: ground handling, climbing, turning, straight-and-level flight, glides and use of the performance instruments.

TECHNIQUE

Brief students on the objectives, transfer control to the students as soon as they are capable, demonstrate each new concept and allow the student to practice, and explain how to achieve desired performance.

DESIRED RESULT

Student can taxi, enter and hold a steady climb, transition to level flight, adjust trim and power, make 90-, 180- and 360-degree medium-bank coordinated turns, maintain straight flight and hold altitude, and enter and hold a steady glide. Added elements; use of radio communication, maintain control during takeoff roll, follow the preflight inspection and use of checklist.

CHAPTER 2

CHANGING PACE

As we progress through the four fundamentals of flight, we find that the airplane must often slow down or speed up, stabilizing at a new speed to maximize the desired performance. There are consequences for failing to control these shifts in speed and, in extreme cases, serious repercussions will occur. Once the student has learned to climb, glide, turn and stay level, he or she must progress to making such changes in speed with precision, and learn to recover from dangerous corners of the flight envelope.

One of the first training maneuvers we will introduce is our old friend slow flight. This seemingly simple task of slowing down, stabilizing, and then speeding up, while maintaining a constant altitude and heading, yields benefits in practically every aspect of flying. It's an excellent tool to sharpen landings, it helps stall recovery become second nature and it teaches students to think ahead of the airplane, a most important skill.

ADJUSTING SPEED

Slow flight is begun from straight-and-level flight. The first action is to reduce power to a setting that will assure a rapid decay in airspeed, but remaining well above idle, thereby retaining some heat in the cylinders and keeping the rate of deceleration more controllable; I use about 1,500 rpm in a Cessna 172 and, for simplicity if equipped, I leave the carburetor heat off unless conditions warrant otherwise. The nose is raised gradually at first, adding just enough back pressure to keep a constant altitude by steadily robbing the energy reserve of excess airspeed. More and more force is required to hold the nose up as the speed deteriorates, unless trim is added; trimming can normally be avoided in small airplanes for the temporary task of slow flight training.

The target airspeed for beginners is usually 10 knots over stall, which avoids triggering the annoyance of a constant stall warning. As we near this speed, power is brought back up to a setting that will exactly maintain altitude at slow flight speed, no less, no more. Don't wait for the airspeed needle to touch the target, because there's a second of response time before thrust builds up to stop the momentum of deceleration. A bit of right rudder will be needed to offset the increase in P-factor so the heading will not change, and often a small adjustment in nose attitude is needed as the propwash increases, if the speed is to remain stable.

The student is now engaged in a balancing act; just enough power is used to maintain altitude, the nose attitude is held steady to keep airspeed on target and enough right rudder is applied to hold the heading. Correct any tendency to make altitude corrections with pitch changes, or speed adjustments with the throttle. Slow flight drill is designed to teach that power primarily controls altitude and airspeed is changed by raising or lowering the nose. There is, of course, an intertwined link between the two, but quick, safe corrections should be taught by controlling speed with pitch, altitude with power.

We can demonstrate this by adding 200 extra RPM, which begins a slow climb, and then reducing power to 200 less than the level-flight number, causing us to sink if airspeed is held constant. In the absence of other available energy, power is obviously the source of altitude. And then we can demonstrate the benefit of flap extension, which usually delays the onset of stall warning; with flap added, the nose attitude will require adjustment and power must be increased to offset the added drag. I normally allow low-time students to practice slow flight without the complication of lowering the flaps. Once the principles of slow flight are learned, changing the airspeed target or configuration is easily accommodated.

Recovery from slow flight normally involves adding full throttle, retracting half of any flaps used, nudging in more right rudder and gradually relaxing back pressure to lower the nose while accelerating, at a rate just fast enough to halt a climb, yet preventing any loss of altitude from lowering the nose too early. We will wind up in hands-off straight-and-level flight, right where we began.

FORCED LANDING EMERGENCIES

Once having mastered the fundamentals and slow flight and experienced a few approaches and landings, the student should be introduced to the concept of handling an emergency involving the engine. Naturally, we stress the reliability of modern powerplants but we also teach risk management, and as a safety measure the student must know what to do if the engine quits. We've already talked about this during our introduction to gliding flight, but now the proper engine-out drill must be taught.

When faced with a loss of power, the first task is to preserve altitude as long as possible, which means we must start trimming the nose up to slow from cruise speed to best-glide speed before we lose even a foot of altitude. Reducing the rate of sink by slowing down buys some time, allowing us to scan the countryside below for an open spot. I point out the fields that are too small, the unfeasibility of landing in a housing area or forest, and the need to find some spot that's level, smooth and into the wind. Once a suitable area is selected, we proceed to troubleshooting, which may obviate the need for the forced landing; fuel flow is confirmed by checking tank selector, mixture, auxiliary pumps and primer, carburetor heat is turned on to bypass the air filter and heat the carburetor throat, and we can simulate switching magnetos off and on, which might help diagnose the problem.

The student with only a few hours can't be expected to fly the landing pattern, but they should be taught the procedure and can follow through on the controls as the forced landing is set up. There's no point in proceeding lower once having turned onto a long final, but the successful outcome should be obvious, as a confidence builder. I then warn the student that they can expect a simulated engine failure on every flight, which they are expected to handle appropriately.

THE ULTIMATE EXTENSION OF SLOW FLIGHT

Slow flight is also useful as a training precursor to stalls. Knowing that many students have heard or read all kinds of horror stories about the wild rides encountered during stall training, I like to begin by placing the airplane in stable slow flight, explaining how increasing the wing's angle of attack has limits, pointing out the wing's obvious angle to the level flow of oncoming air. Eventually, I tell them, as this angle is increased the air can no longer make the sharp corner over the leading edge, and it will break away in burbles instead of maintaining a smooth flow, thereby disrupting lift. Only the inboard portion of a properly-designed wing will stall at first, so the airplane does not enter a free-fall; it simply sinks due to insufficient lift. Because the stall is progressive, spanwise, some lift remains near the wingtips.

By now, we're right at the stall, so I pull the nose a little higher, induce a stall shudder and point out that sink rate is beginning to show on the VSI. I pop the stick forward to lower the angle of attack slightly, and the airplane is once again in stable slow flight. "Now, that wasn't any big deal, was it?" I reassure the student. The point I'm making is, if the pilot recognizes that a stall is taking place, takes immediate corrective action and returns to flight with a proper angle of attack, the only penalty will be a hundred-foot loss of altitude.

Fear of stalling is healthy; that's what keeps us alive and well, encouraging us to take precautions to avoid stalls at unwise moments in the flight, such as during takeoff and landing. Panic during stall recovery, on the other hand, has no place in the pilot's mind. If aircraft control is given over to panic, the pilot will revert to primeval instinct, clutch the wheel full back to avoid falling, and ride an out-of-control airplane to the ground, all the while just a yoke-push away from escape. Learned behavior must replace instincts, achieved by lengthy practice in stall recovery from all possible conditions of flight.

With the ice broken, I explain that adding full power reduces the altitude needed to recover from the stall, and that we will never practice stalls at an altitude less than sufficient to recover above a minimum of 1,500 feet AGL. I reassure the student that a properly-trimmed airplane will recover to un-stalled flight on its own, just by releasing the controls, albeit with 500 feet or so of altitude loss. The student should then have no fear of stalls, but can simply approach them as another maneuver to be mastered.

ONE STALL LEADS TO ANOTHER

I begin stall practice by teaching how to do a simple power-off stall. Idle power eliminates torque complications in the stall, so it's nothing more than slow flight taken to the extreme. With throttle pulled back, the nose is raised to the familiar climb attitude and held there with increasing back pressure. The stall warning comes on, the airframe shudders and the nose breaks downward as the tail stalls, signaling loss of control. Quick application of forward yoke movement and full throttle brings the airplane back to straight-and-level flight. The student must understand that it takes a second or two of acceleration to get the airspeed back into the green, so the stick or wheel must not be yanked back quickly to level off, but left forward momentarily to avoid a secondary stall entry.

The student's practice of a few of these innocuous stall excursions builds confidence in the process, and then we can demonstrate a power-on stall, often called the takeoff stall. The familiar entry to slow flight is performed, then full power is added to simulate the post-takeoff climb. The nose is raised above the usual climb attitude, thereby assuring airspeed will fall off as the angle of attack steadily increases. The importance of adding right rudder to maintain straight flight is stressed, by showing that the nose drifts to the left if not corrected, slowly pulling the left wing down. The power-on stall is decidedly more sudden and pronounced, requiring quick recovery action, but it's easier to evade because power is already in place, so the only thing needed is a push on the yoke and perhaps some rudder to prevent a turn. Once angle of attack is lowered, the power-on stall is history.

If the student is apprehensive of this more violent stall, they can be allowed to perform the first ones with reduced power, using the slow-flight setting or perhaps low-cruise power. Eventually, we can work up to full-throttle climb power. Once the need for P-factor correction is understood, we can proceed to the takeoff-and-departure stall, which introduces the notion of stalling while in a departure turn, climbing out of the traffic pattern. The procedure is the same; reduce power, slow to takeoff speed in level flight, re-apply full throttle, then enter a shallow-bank climbing turn. Only when this turn is stabilized—that is, the bank is not changing—will we begin increasing angle of attack to induce a stall. Bank is kept constant, right rudder is applied to counter P-factor (regardless of the direction of the turn, it must be pointed out) and the rate of deceleration is kept low, so as to avoid extreme nose-up attitudes that might precipitate a whip-stall or tailslide.

The takeoff and departure stall is usually manifested by sudden loss of control, often departing in a roll toward the higher wing while the nose breaks downward. This is serious stuff, requiring prompt action on several fronts; moving the stick forward to break the stall while adding opposite rudder to halt the turn, followed by aileron use to raise the lowered wing as speed begins to build. There's no reason to allow the out-of-control bank to reach 45 degrees or the nose to go down 30 degrees below the horizon. Immediate, aggressive recovery action should return the airplane to accelerating level flight before these reactions take place.

At the other extreme, some students are reluctant to move the stick forward enough to fully recover, or will yank the nose back to level attitude prematurely, revisiting the stall. Once the stall warning horn shuts off, it should not come back on. Most students can either break the stall or stop the turn, but not both at once, until they've practiced the recovery a few times.

Caution students to never have their mind made up about what's going to take place in a departure stall (power-on stall). Take action only after the airplane's intentions become clear. If bank is allowed to increase during the entry, the stall break may be hidden in a mushing, semi-stable spiral descent; keep the turn shallow, as it should be for a departure climb. Perform a recovery at any time the stick reaches the full-back position during a stall entry, even there's little or no evidence of a stall.

If insufficient right rudder is used during an entry to a left-hand departure stall, P-factor may precipitate a stall break toward the low wing, instead of the straight-ahead or over-the-top break of a coordinated turn. If the stall entry is done with no rudder applied, feet flat on the floor, the aileron will usually wind up being held against the turn in an attempt to keep the bank shallow. The slip ball will be displaced to the outside of the turn, and that's a perfect set-up for a spin entry. The CFI should be ready to perform an instantaneous spin recovery; throttle idle, opposite rudder, stick forward, and then pull out of the resulting dive. Once this is experienced, most students will diligently avoid ever performing a cross-control stall.

I am frequently asked, "do you teach spins?" My response will be, "no, I teach not spinning." It is important for a student to know what a spin is, and how to prevent it, but intentional spins are beyond the scope of primary training. Recovery from the insipient phase of a spin is a routine occurrence when teaching departure stalls, because students will occasionally stall in uncoordinated flight. If instructors are not confident in their ability to recover from a full spin, they should get refresher training from an instructor who is confident in teaching spin training.

AN APPROACH GONE AWRY

As a break from the potentially-violent takeoff and departure stalls, the innocuous but equally dangerous approach to landing stall is introduced. This is a complex maneuver, but, like the departure stall, the entry must not be rushed. Each step is important and should be given due attention. The approach to landing stall is actually a full course in flying the airplane, compressed into a few seconds of feverish concentration. Everything learned to date will be used: flight controls, trim, carb heat, flaps, airspeed, throttle and multiple attitude references.

Entry is begun by performing the landing sequence: carb heat on, power to idle, trim adjusted while speed is reduced to the flap operating range by maintaining level flight attitude. Once airspeed is in the white arc, flaps are extended and the nose is lowered to maintain normal approach speed. Final flap setting is made and the pull-up to the stall is begun; simply holding a steady altitude until reaching the stall break is a good technique. The normal approach stall is fairly benign, particularly at forward CG locations. As full-up elevator is reached, the nose bobs up and down as the airplane repeatedly stalls and un-stalls itself, with little tendency to roll off on a wing.

Recovery takes some effort. Moving the stick forward breaks the stall, full power limits altitude lost (don't forget to push off carburetor heat), right rudder nullifies P-factor, and then, as airspeed builds, an additional shove on the yoke may be needed to prevent the trim from re-stalling the airplane. Most students immediately reach for the trim wheel, which calls for a ruler applied to the knuckles; the flaps should be retracted before trimming, an action that will cancel out much of the mis-trim force. As the flaps come up, the nose will need to be raised to keep airspeed at about 1.3-times clean stall speed; the plane should be transitioning into a climb, away from the simulated runway elevation. Remember, we're replicating a stall on final, at very low altitude; minimum altitude loss while cleaning up the airplane is important.

Attitude control is critical during the recovery. To assure immediate exit from the stall, the nose must start out well below level flight, but not left there too long, in order to limit altitude loss. Only a recovery to landing-approach speed is needed; excess airspeed means altitude is being dived away in a high-drag configuration. Once safely out of the stall, the nose is raised to about level flight attitude, but no higher, where the airplane can fly level in slow flight with flaps extended. After confirming airspeed in the green arc, the flaps can be brought up, at which point the nose is raised to a climb attitude, replacing the lost lift of the flaps with increased angle of attack. These three distinct nose attitudes must be flown appropriately at each stage of the recovery.

Students learn the sequence of events quickly, and will soon start slamming and slapping throttle, trim, flaps, and carb heat like an automaton. This causes events to exceed requirements; flaps can't be fully retracted until airspeed has increased, trim doesn't need to be adjusted until the flaps have been cycled, nose attitude must be adjusted to fly the proper profile to clear the ersatz trees. Fly the airplane, I preach, don't just react.

To challenge the student further, the approach stall should also be performed while making a simulated turn to final, thus requiring added control inputs in the recovery to return to straight flight. Limiting bank to normal approach turn angles is necessary to create an easily-detected stall break. Carrying a partial-power setting into the stall and/or using approach flaps instead of full extension can also add some excitement to what is usually a fairly tame stall.

The value of approach stall practice is that it requires manipulation of many controls and flight conditions in a short amount of time. It is a landing in miniature, followed by a go-around, with much flying required. Mastering this complex beast pays big dividends in perfecting traffic pattern technique.

Other stalls that can be introduced are a demonstration of the trim stall, which is basically a hands-off roller-coaster ride following the application of throttle while trimmed for a landing approach, and the imminent stall recovery, which teaches students to recover at the very first nibble of a stall instead of waiting for the airplane to deliver a summons. One can also show advanced individuals an accelerated stall, which induces a stall at higher than normal airspeed by placing added g-load on the wings, normally demonstrated by holding altitude with reduced power in a steep bank.

Students should become so familiar with stalls that they treat them as normal parts of the airplane's flight envelope, to be avoided except when practiced at a safe altitude, but nothing to be feared. Every landing, after all, requires passing through the stall regime.

LESSON PLAN

OBJECTIVE

To acquaint the student with the proper way to control the airplane during airspeed changes, including stall recognition and recovery. Student will not fear stall practice under safe, controlled conditions, but will be ready to avoid and recover from a stall at any time.

TECHNIQUE

Brief the student on the objectives, stressing the safety benefit of learning to recover from critical flight conditions. Demonstrate slow flight entry, maneuvering and recovery, followed by directed practice. Build proficiency in stall entry and recovery by first using simple power-off, clean configuration, straight-ahead stalls, then adding the complexity of power, flaps and bank as the student becomes proficient in each type of stall.

DESIRED RESULT

Student can stabilize the aircraft in slow flight at a designated speed and configuration without allowing uncorrected altitude or heading errors, recovering to straight and level flight on the beginning altitude and heading. Student enters and recovers from each type of stall unaided, achieving consistent performance and exhibiting the skill and confidence to practice them solo.

CHAPTER 3

TRACING A PATH

It's not enough to steer an airplane around the sky, ignoring its path across the ground. The purpose of flying is to make the plane go somewhere, often along a designated course, rather than just allowing it to wander aimlessly. With fundamental flying and speed control learned, students must now start looking under the airplane and maneuvering in relation to ground references.

The seemingly pointless tracking of roads and hedgerows, or making loops around specific targets, doesn't fit into everyday flying—does it? Ah, but the skills acquired by these inane antics are vital to a pilot's development. Performing the airport traffic pattern, at least part of which is going to be traced on every flight, has its roots in the rectangular course we'll teach in ground reference training. And who knows when a pilot might need to make a circle around an objective on the ground, such as an airport windsock or a checkpoint-confirming water tower?

Given my druthers, I would prefer to teach ground reference maneuvers on a day with a 10-mph wind at traffic pattern height. A calm day requires simulation of wind drift angles and is useful only for perfecting control of altitude and turn anticipation. But with the proper amount of wind, the student can readily see why corrective action is required.

We begin by commanding a level-off at an unfamiliar 1,000 feet AGL, not the higher level required for slow flight and stalls. By bringing the airplane down to near traffic pattern altitude, the student can better observe drift and detail that would otherwise be lost from above in the haze. The goal is to make the world appear as it does when maneuvering around the airport. Now, for the first time, the student is having to divide their attention between the world outside and the dashboard inside, scanning the altimeter, watching the roads, and always remembering which way the wind is blowing.

CHASING CRABS

Let us begin by tracking along a straight road that happens to lie perpendicular to today's wind. Don't fly directly over the road, but slightly to the right of it, so it's in clear view of the person in the left seat. Observe that when the airplane is aligned with that road, the roadway begins to move laterally in relation to the airplane's longitudinal axis, as the gap widens or narrows from the effect of the crosswind. Clearly, the airplane is drifting with the movement of the air in which it is flying, like a log floating down the river. If we want to arrive at a boat dock on the opposite shore of this free-flowing river, we must angle the prow of our craft upstream, fighting the current with a crab angle. Crabs walk sideways along the beach, and so must we "crab" in a crosswind.

The student must understand that it's not necessary to continually hold pressure on the controls to do this, forcing the airplane to maintain the crab angle. Instead, we simply point the nose of the airplane a few degrees into the wind and neutralize controls to fly straight and level, thereby maintaining a straight track along the road while pointed to one side. The airspeed reads a normal cruise number. As we reach the next crossroad, we fly around the corner and head directly into the wind. No crab angle is required, but we've visibly slowed from the headwind now impeding our progress. The airspeed indication, however, remains unchanged; obviously, groundspeed changes resulting from the wind's effect do not register on the airspeed indicator, which merely shows our speed through the air, even though that air is moving across the earth.

At the next junction, we swing around to once again parallel the road or fence line, this time angling to the other side of the road to set up our wind crab. This keeps our ground track aligned until we reach another turn point, where we'll head downwind. Now we're really scooting across the ground, but once again the indicated airspeed is unchanged. It's time for a test; I'll ask the student, "is this next turn going to be 90 degrees, less than 90, or more than 90?" The attentive ones respond, "more than 90 degrees," realizing they must anticipate the need for an added crab angle.

Well and good. We've proven that one must correct the airplane's heading to nullify a crosswind's drift, and that airspeed is unaffected by head or tail winds. We must now see how wind affects the aircraft's ground path during a turn. We head into the wind to fly over a prominent landmark and announce our intention to start a 360-degree turn upon reaching that point. I roll into 30 degrees of bank, hold it perfectly steady all the way around and, upon rolling out, the student can see that we've drifted downwind during the turn, and consequently we're nowhere near our starting point. We struggle back to it to start over.

This won't do, I insist. Imagine there's a nudist colony I wish to observe in the center of the circle, so I must somehow correct the offensive wind to keep the sunbathers in sight. This time, I begin with a 15-degree bank, which, but for the wind, would create a larger radius of turn. The wind pushes us back as we start around the circle, so our turn radius is kept to the same distance as it would be with a 30-degree bank on a no-wind day. As we approach a point on the circle about 90 degrees from the beginning, the shallow bank is no longer useful and we must steepen to a medium bank, swinging around to a downwind heading, where we must employ a steep bank of perhaps 45 degrees in order to stay on our desired perfect track. With no wind, the steep bank would tighten up the radius of our circle, but with today's wind pushing us away from the point we're circling, we remain on track. The steepened bank requires added back pressure on the yoke to maintain altitude, but the faster rate of turn means we'll only be in the steep bank for a little while. After passing a heading perpendicular to the wind, we return to a medium bank and, lo and behold, we're right over the spot where the circle was begun.

The student now must fly their rectangular course with not only crab angle adjustments to the crosswind legs, but with steep or shallow banks when turning to crosswind, maintaining equal size turn radius at the corners. Altitude is to be held constant, all this while. We've now planted the seeds of traffic pattern flying.

THE GRAND SLALOM

When briefing this lesson, we also sketched the diagram of S-turns across a road, a fiendish little maneuver that has much to teach student pilots just beginning to look outside while flying. The concept is simple; find a road running perpendicular to the wind, head across it with the wind at your back, and make a 180-degree turn so as to cross the road exactly as you roll out, immediately entering a 180-degree turn in the opposite direction, with the same objective. And so it goes, back and forth, swooping across the road in perfect half-circle tracks, despite the wind's best efforts to thwart our intent.

From the foregoing practice, the student knows a steep bank will be needed on the downwind side of the road, and a shallow one on the upwind side. There can be no hesitation as we cross the road; aileron and rudder are applied with enthusiasm because the steep bank is needed right away. But, halfway through the 180-degree turn, we have to shallow into a medium bank because we no longer have the tailwind chasing us away from the road. The important rule of making good S-turns across a road is, "the bank cannot remain constant throughout the 180-degree turn." Hopefully, the steep bank had its desired effect and our medium bank can serve to carry us around the rest of the turn to reach our roll-out point exactly over the road.

However, more than likely we'll need to watch our progress toward the road and make alterations in our bank, to speed up or slow down the rate of turn, in order to use up the remaining distance to the road while our heading comes around to the 180-degree point. The student is thereby learning to think ahead of the airplane, to reach a goal by using the controls in anticipation of need. Playing catch-up with an airplane is not piloting, it's merely being a passenger in the left seat.

As we cross the road and roll the wings level, we continue rolling into the opposite turn; the airplane should never remain wings-level in a well-flown series of S-turns across a road. Patience is required on the upwind side of the road, because the shallow bank required initially means a protracted interval passes while swinging away from the road. Most students can't resist the temptation to hurry things along, steepening the bank prematurely and running themselves out of room to finish the turn, because they've drifted in toward the road. They must be cautioned to stay in the shallow bank until nearing the 90-degree point, beginning the roll into a medium bank while still crabbed into the wind, at the apex of the loop across the ground. With cunning and skill, the space remaining will be sufficient to finish the 180-degree turn and allow a rollout just as we cross the road, and we'll continue into another S-turn without hesitation.

As I always say, when flying S-turns across a road, "keep one eye on the road, one eye on the altitude, and one eye on the bank." If there are inevitable mistakes made, no worries; just re-enter over the road and avoid making the same mistake again. S-turns always give you a chance to make a fresh start. I like to throw a change-up pitch to students after a half-dozen S-turns by reversing the direction of progress along the road. This requires a mental switch because a steep left turn becomes a steep right turn, and vice versa.

WHAT IF...

It's important to include forced-landing drill while flying close the ground. When we first discussed engine failure and how to handle it, we were flying at an altitude that gave a lengthy amount of deliberation time. However, a low altitude engine-out emergency doesn't leave much time before a landing will be taking place, one way or another. I usually pull the throttle closed when flying crosswind, or even downwind, to see if the student will set up an approach into the wind or try for a perfect spot that happens to oriented downwind.

The important thing in low-altitude emergencies is to react speedily, in the correct order. First, prolong the time by holding level flight while slowing to best-glide speed, trimming as you slow so the airplane will stay in the glide while you're busy elsewhere. Check immediately for landing spots in an upwind direction; you might be leaving the only suitable place in the area. At least head for an open patch of ground, while there's time. With the plane under control and options under consideration, check the vitals; fuel tanks switched, pump on, mixture adjusted, carb heat on, throttle setting changed, magnetos cycled. If nothing works, you're already set up for the landing, by planning to land into the wind.

Beginning at 1,000 feet AGL, have a field picked out within reach, and when at 800 feet, be in the landing pattern, and by no less than 500 feet, be turning onto a base leg, preparing for the landing. In reality, a low-altitude forced landing is easier to perform, because there are fewer options to consider. One simply reacts to make the best of a rapidly-developing situation, and the deed is quickly done. A high-altitude emergency, on the other hand, allows time for vacillation, which a low-time pilot will take full advantage of, choosing first this field, then that one, or perhaps forgetting to troubleshoot on the way down.

Ground reference flying provides valuable training in thinking ahead of the airplane, visualizing where the wind is blowing in relation to the plane and dividing one's attention to successfully multi-task. When a student is struggling to keep up in the traffic pattern, going back out to the practice area for some ground-reference work can be helpful.

LESSON PLAN

OBJECTIVE

To teach the student how to maneuver the aircraft in relation to ground objects, by correcting for the adverse effect of wind drift and dividing attention between the dual tasks of flying the airplane and following a ground track.

TECHNIQUE

Brief the student, graphically and verbally, on the maneuvers to be flown and how to offset wind effect. Demonstrate the effect of wind on the airplane's ground track, then show the proper corrections to negate the wind drift. Direct the student's practice of the rectangular pattern and S-turns across a road.

DESIRED RESULT

Student will be able to perform rectangular courses, similar to traffic patterns, while maintaining straight ground tracks and holding altitude within 100 feet. They should be able to fly S-turns across a road without prompting by the CFI, making equal-size semi-circles and reversing direction as the road is crossed at the 180-degree point. Student should eventually be able to practice the maneuvers solo.

CHAPTER 4

GAUGES AND GADGETS

One of the greatest challenges to a person learning to fly will be the plethora of gadgetry on the instrument panel. Some things are somewhat familiar, such as the tachometer, oil pressure gauge and fuel gauges, but even these have their own quirks to be learned. With a fixed-pitch propeller, the tachometer needle doesn't hold a steady indication as airspeed increases or decreases. The oil pressure may only have a green arc to indicate a normal operating range, and fuel quantity, we're told, must be verified by an independent opinion, instead of trusting the gauges. What gives?

It will take some clear and careful tutoring to explain the characteristics of each instrument on the panel. The magnetic compass, for instance, is probably hidden away in the windshield, and it looks nothing like the round-dial Boy or Girl Scout compass the student might have encountered in the past. There's a similar gadget in the middle of the panel, called a directional gyro (DG), and its purpose, which is to supplement the magnetic compass, has to be demonstrated by showing how it tracks smoothly around a turn while the compass card wobbles, leads and lags. Because the DG is only a mechanical device, it must be periodically updated from the magnetic compass during stable flight, as it tends to forget where it is over time.

Each time I watch my student set the directional gyro, I make them tell me the heading in degrees that is being shown by the compass. Then I cover up the compass and tell them to set the gyro to that heading, without looking back and forth to the compass reading. If I don't do this, they will not only never learn how to properly report their heading to ATC, they will invariably set the gyro's vertical card to resemble the horizontal compass card, placing "E" on the left side of the lubber line when the heading is 105 degrees, which sets the gyro to 75 degrees, a 30-degree error. "Read the heading off the compass, and never look back until you've finished setting the gyro!" I stress.

Setting power to an RPM indication can be frustrating, because the tachometer reading slows down with an airspeed reduction and speeds up if airspeed increases, as frequently happens when the student tries to hold an exact altitude in rough air. This coupling between airspeed and engine rpm, as the engine labors in a downdraft and races in an updraft, leads to distracting throttle-jockeying until the student learns to set-and-forget unless a major altitude change is required.

Fixation on an instrument indication frequently interferes with learning to actually *fly*, rather than merely drive about the sky. As we attempt to foster a desire for precision, students are frequently chastised for allowing the airspeed to slip off the target number, or letting the altitude or heading wander. This leads to fixation on the performance indicator, as if staring at the needle and moving the controls to change it will serve to guide the aircraft precisely. It must be pointed out, rather continually, that establishing the correct attitude and power is the way to produce the desired results, rather than just "flying the needle." The inherent lag between aircraft movement and change on the instrument face means a "needle watcher" will fly a constant series of corrections, until settling on a power and attitude that works.

It's better, then, to fly attitude first, using outside references as much as possible, and check the performance instrument, be it altitude or airspeed or rate of climb, to grade the effort. I sometimes cover up the instrument panel with an unfolded sectional chart and demand that a turn, climb or descent be flown with only the natural horizon, then lift the sectional periodically to show that the technique works. Hopefully, the student will apply this skill, instead of "flying the needle."

ON THE GAUGES

The FAA requires that students be given training in emergency instrument flying prior to solo cross-country sign-off, so there's no reason to wait until late in the syllabus to introduce blind flight. Because most beginning students have stared at the instruments from their first hour, including the bank index on the artificial horizon, putting on the instrument flying hood is not a great challenge. However, it's important to stress the need to scan the appropriate instruments, because no one gauge will give all the information needed. The 3-inch artificial horizon is a poor, albeit usable, substitute for the 3-foot one above the glareshield; it needs the directional gyro, altimeter, airspeed and turn needle/slip ball to support it.

Attitude instrument flying, like the basic fundamentals taught in hour one, integrates the aircraft's attitude and power settings to produce results. As I said, there's nothing new about this, as far as the student's concerned. But the danger of relying on one's internal feelings, when there are no outside visual references, must be understood. An untrained individual, ignoring the instrument indications, will gradually initiate a "graveyard spiral," an ever-tightening turn with increasing speed and g-load, usually within 45 seconds, with fatal results. The student must ignore the visceral and inner-ear sensations and focus on the instruments, making them substitute for the missing visual references.

There are three requirements for successful instrument flight: control of the aircraft by reference to the instruments only, spatial orientation with the aid of electronic bearings, and communication and compliance with ATC. Many pilots can perform one or perhaps two of these criteria, but not all three at the same time, a juggling act that's vital for survival in the clouds. For the low-time student pilot it's best to teach aircraft control as the primary task, adding orientation and radio communication as supplemental goals to be pursued as subordinate aims.

Instrument work begins with holding a heading and altitude, simple straight-and-level flight, by making the attitude indicator the hub of a wheel with spokes radiating out to the supporting instruments, in this case the heading and altimeter. Then, a turn to a specific heading is made, introducing the concept of a standard-rate turn, established by holding to a bank angle that keeps the turn needle nestled against the standard-rate index. The scan must then include the rotating card of the directional gyro, with rollout begun five degrees or so before the desired heading. However, the rollout maneuver is best performed by watching the bank index of the attitude indicator, stopping it exactly in neutral, instead of trying to arrest the motion of the DG.

Responding to nervousness and panic, beginning instrument pilots will tend to apply back pressure on the controls, climbing and slowing unconsciously. To minimize this, they need to recognize that this unwanted back pressure is normal and to watch for it, countering it by periodically relaxing and releasing the controls. I suggest wiping the flying hand on the pants leg, or laying the non-flying hand across the flying wrist, for calming influence.

Climbs and descents are simple extensions of what's already been learned visually; to go up, the nose is raised to the correct pitch attitude by watching the horizon bar of the artificial horizon, power is applied to maximize climb rate, and trim is adjusted as needed. Leveling off is done by first lowering the nose to the level flight position on the attitude indicator, then making a cruise power adjustment as the plane accelerates. I like to have students adjust cruise power by ear, while keeping attention on the flight instruments, instead of watching the tachometer and letting the airplane wander; they can then check the tach to see how close they came. Most of the time, the power setting is correct, because they've already learned the sound of normal power.

Making a descent through clouds is a vital survival skill for non-instrument pilots. Descents are begun by reducing power to a setting that will produce about a 500-fpm rate of descent; no trim should be needed, because we are not changing airspeed. There will be a slight nose-down adjustment in attitude on the artificial horizon; most airplanes will initially want to pitch down excessively until stabilizing in the descent, motivating low-time pilots to keep the nose up instead of adjusting to the new, lower pitch attitude. If a turn is needed, the descent should be established first, rather than trying to perform two tasks at once. Leveling off requires only a return to normal cruise power; the nose should rise to level flight attitude on its own, if trim wasn't adjusted.

The other critical survival skill for inadvertent instrument flight is the 180-degree turn. If the weather was better where you came from, you should obviously turn around and go back. On the modern directional gyro instrument, the reverse heading is right there on the bottom on the gauge, upside down and opposite the current heading. Note it and roll into a standard rate turn, and scan the instruments while the reciprocal slowly rotates to the top of the instrument. One minute after entering the turn, roll out on the heading that will return you to good weather. If you can perform this maneuver calmly and under control, you will survive to fly another day, wiser than you were.

Learning to use the gauges and gadgets in the cockpit occupies a lot of a student's time, until their purpose becomes clear and terminology is second nature. A student must never be reticent about clarifying a confusing term or reference. Always tell them, "if you don't know what the CFI is talking about, ask; don't guess." Aviation-speak is not a first language for any of us.

LESSON PLAN

OBJECTIVE

To gain an understanding of how each instrument on the airplane's panel is to be used by the pilot, including its limitations. Adequate precision is to be obtained by outside visual references whenever possible, instead of fixating on the gauges.

TECHNIQUE

Instructor will demonstrate compass errors, airspeed indicator lag, and slip ball and attitude indicator uses. Prior to the first solo flight, the student will be introduced to the instrument flying view-limiting device (hood), to gain fundamental ability to extricate themselves from inadvertent IMC encounters. This should include proper instrument scanning technique.

DESIRED RESULT

Student will correctly respond to questions about the instruments to verify knowledge of their purpose. Basic instrument flying with vision restricted should show the ability to hold a heading, turn to a desired heading, and climb or descend under control. Further development of instrument flying skills will be deferred until later in the syllabus.

CHAPTER 5

CIRCUITS AND NON-BUMPS

Now that the student has learned all the fundamental elements of operating an airplane—turns, climbs, glides, speed adjustment and maneuvering by ground reference—it's time to put all this knowledge to work by employing it to perform takeoffs and landings. Flying around the airport by following a standard traffic pattern involves using everything that's been learned to date, put together in a rapid sequence. This can be a little intense, so if the student has trouble keeping up with the airplane in the pattern, it's time to go back out to the practice area for some review of the particular shortcoming.

During previous lessons, the student has gradually taken over more and more of the flight's departure and arrival procedures, beginning with handling most of the takeoff run. The landing itself has remained the responsibility of the instructor, although each arrival should have been used as a teaching tool, so the student can develop an idea of what's to come.

If the training operation is based at a tower-controlled field, it probably would be best to introduce traffic pattern work at a quiet uncontrolled airport, where there's more likelihood of making a complete circuit with no interruptions. Traffic patterns at such airports are typically smaller, allowing more concentrated instruction, as opposed to flying a two-mile final behind other aircraft. However, if the training must be conducted under ATC supervision, it may be possible to schedule the sessions at less-busy times, or to use an alternate runway.

START SIMPLE

I prefer to introduce flight in the traffic pattern as a maneuver in itself, apart from rolling down the runway. Therefore, we will return from the practice area part-way through a lesson period, and will first fly a rectangular pattern around the field rather than make an approach to the runway. This establishes the proper spacing from the runway, as seen from the cockpit, and it evaluates the wind drift component of the day and lets me point out elements like runway markings, wind indicators and VASI lights.

After a trip or two around the rectangle, it's time to incorporate the vertical dimension into the pattern. The pre-landing checklist is performed on the downwind leg and power is reduced for the approach when we're abeam the landing spot, typically the end of the runway. If practical, given the airplane characteristics and the day's conditions, I like to have the student pull the throttle to idle and fly a power-off approach, to reinforce the concept that it's possible, even normal, to land without power, and that one should keep the pattern small enough that's always possible to make the runway. Having said that, the student should understand that such a goal is only an initial basis for establishing the landing approach, and correcting an approach that has gone low by adding throttle is perfectly normal procedure, as long as the engine is still running.

Some airplanes, and some training situations, will require a partial-power landing approach, in which case the throttle is reduced to whatever power setting is applicable. The goal should be a progressive reduction of power as the runway is neared, with no reapplication of throttle once it has been taken away. Jockeying the throttle back and forth is a sign of poor judgment of the approach and of attempting to fly the engine instead of the wing. As we progress into landing training, some pilots need a wider pattern, with the concurrent longer final approach, to get enough time to study and absorb the lessons of the task. Instructors need to be flexible in their technique.

Once power is reduced, the aircraft is slowed to gliding speed by first maintaining a short period of level flight, while trim is set to hold the desired hands-off approach speed, and attention is given to the runway disappearing over the student's shoulder. A 45-degree relationship, from runway end to cockpit, should mark the key position at which to turn onto base leg. It's impossible to avoid letting the student pick up on the surrounding streets, buildings and fencerows that delineate the traffic pattern, but it has to be stressed that one will not have these familiar landmarks at other airports, so the basic 45-degree position from the runway is fundamental to turning onto base leg.

When initially teaching landing approaches, I do not require rigid, programmed timing of flap, speed and power adjustments. Let's keep it simple; idle the engine, establish a proper glide speed and fly three straight legs to reach the runway. This is the quickest way to learn glidepath judgment, by starting with the bare essentials. Power can be added to correct an approach that's gone low and flaps may be extended to mend a high approach; otherwise, if it's not broke, don't fix it. Most students will not be able to tell what correction is needed until they are on final, so instructors will have to hold their tongue, even when the situation is obvious to them on base leg, allowing the student to make their own tardy correction until they acquire more judgment skill.

ALWAYS FLY THE PLANE

I realize that official policy dictates the making of pattern advisory radio calls on each leg of an uncontrolled-airport pattern. However, it's also important to fly the airplane *first* and make radio transmissions when there's time. Intentions should be made clear by making the call early on the downwind leg, and if the pace of instruction does not allow one of the subsequent calls to be made, defer it until later, in the absence of a traffic conflict. Don't drop the airplane to fly the microphone, but be aware of other traffic in and around the airport.

Base leg is the time to consider lowering any flap needed for the landing. However, if the airplane is capable of landing flaps-up and the approach is going fine without adding flaps, why spoil things just for procedure's sake? The student must learn that power and flap adjustments are tools for modifying a landing approach, not the sole means of flying the airplane. There's no point, for instance, in lowering approach flaps on the downwind leg unless you're flying a high-performance ship that needs to slow down for spacing. Such procedures will be taught later, after fundamental traffic pattern work is learned.

The turn to base leg should be done with a medium bank, to keep the pattern tight and limit wind drift effect. Crab angle should be included to make a straight rectangular ground track, and the turn to final anticipated so it can be safely done with less bank than was used in the turn to base leg. As the runway slides into position over the nose, airspeed is rechecked for correctness and any need for glidepath adjustment becomes apparent. If the runway tends to move up in the windshield, the approach is becoming too low and extra power is needed. How much? My answer is, enough to feel the added thrust, as an upward surge under your seat. More power is always better than too little, but small early corrections are more desirable than late, massive ones.

If the runway is dropping out of sight below the engine cowling, the approach is becoming too high, and making the landing in the first third of the runway will require steepening the approach, by using less power or more drag, perhaps both. Any throttle remaining should be closed and about half the available flaps extended, which will require attitude and trim adjustment. Save full flaps for later, unless the situation warrants their use right away.

CROSSWIND COMPLICATIONS

Some students get all the luck; they will encounter crosswind conditions every time they need traffic pattern work. Be ready to introduce the sideslip early on, even though low-time students will be overwhelmed by adding an extra task to their landing approach. The alternative is to line up on final with a crab angle, explaining that it will have to be "kicked out" to align the tires with the runway at some point.

I demonstrate crosswind correction over a long runway, taking off and then reducing power to "hover" along the centerline. With no correction, the airplane drifts downwind, off the runway, and if aileron is used to get back over the pavement, the nose will no longer be aligned with the center stripe. Then I show the student how kicking in opposing rudder puts the longitudinal axis back where it belongs, while holding upwind aileron creates the sideslip that prevents drift. Having some flaps extended improves over the nose visibility for the demonstration, concluding with a cleaned-up crabbing climb-out to the traffic pattern, where we set up another approach.

PLAN TO GO AROUND

When we reach the runway, we will turn the glide into a climb by making a planned go-around. At this point, we are introducing the student to the traffic pattern, and making a good approach is the start of making good landings, so we'll conduct wave-offs instead of full landings the first few times. In this way, the student learns that landings are optional and that it's always possible to go around and try again, without penalty.

The trim and attitude changes caused by an application of go-around power need to be learned, while flaps are retracted and airspeed stabilized at V_Y by raising the nose. Typically, students will allow the speed to creep up toward cruise numbers unless an effort is made to teach a proper climb transition. A radio call, advising of the go-around, comes only after the plane is established in the climb. The departure path may be adjusted to one side of the runway centerline to provide a view of potential traffic below.

After climbing straight out to 400 or 500 feet, or 200 below the pattern altitude, or whatever local procedure works best for the airport being used, the turn to a crosswind leg is made and a point is chosen for turning downwind. Never make a continuous 180-degree turn from climb-out to downwind; always roll out at the 90-degree point to check both sides of the aircraft for traffic, then turn onto the downwind leg. Level off exactly at pattern altitude, reduce power to cruise, set trim for level flight and make an early advisory call. Repeat as necessary, until the student is capable of keeping up with the pace of the traffic pattern. Then, we can proceed to actual takeoffs and landings.

LESSON PLAN

OBJECTIVE

Integrating all the previously learned elements into a complete traffic pattern, learning to make a safe approach, go-around and climb-out.

TECHNIQUE

Begin with a two-dimensional rectangular pattern around the airport, then add the descent to the runway and climb-out to pattern altitude. Show how to adjust glidepath to bring the aircraft to the desired touchdown zone. Avoid complex procedures until the basics of a landing approach are learned.

DESIRED RESULT

Student will establish the proper size traffic pattern and pattern altitude, make prelanding checks, enter and hold correct glide speed, and initiate corrections to adjust glidepath. A safe go-around transition to climb back to pattern altitude will be made, using correct traffic avoidance procedures.

CHAPTER 6

THE ART OF ARRIVAL

Almost without exception, every student will at some point declare, "I can do everything else, but I have trouble with the landing." In response to which, I assure them that no aircraft in history has ever remained aloft forever, so they shouldn't be concerned. Airplanes always come down.

If a student's landings truly aren't going well, their foregoing statement about being able to handle everything else is probably in error. Landing problems will invariably be related to fundamental mistakes in basic flying that need to be corrected. The airplane lands on its own, if it's simply flown to the point of skimming over the runway and decelerating toward the stall. Quite often, it's the approach that is the source of what's perceived as a landing problem. Or, the mental distraction of the earth's surface rapidly drawing near can cause vital skills to suddenly atrophy.

Once the student learns how to combine the fundamentals of flight into a fluid string, as must be done to fly the traffic pattern, it should be possible to make a landing. They know how to place the plane into a steady glide, and to combine a turn with the glide, and to perform a transition from a glide into slow flight. The only trick, then, is to do these things as the runway draws closer to the wheels. Ah, but it's much easier to talk about it than to do it...

When practicing the basics of flying at altitude, there's a couple hundred feet of room to move around in, perceptible only through deviations shown by the altimeter or compass. However, the runway is a target only 50 feet or so in width, with a finite length available for stopping, and there's a constant concern about stalling out too high and dropping in, making for a hard landing. When the ground comes up, the pressure's on. It's no wonder students worry to excess about being able to land on their own.

START BY TAKING OFF

Oddly enough, learning to land begins with making good takeoffs. When you have the student pull out onto the runway and line up for the takeoff run, they are looking at a picture of how a landing should end. The goal is to be on the centerline, an equal distance from each edge of the pavement, with the aircraft's longitudinal axis aligned with the runway. Just before touchdown, the nose must be raised to a slow-flight attitude, and the runway edges should be visible beside each shoulder, not far below your elbow, but just a little lower than where they appear as one prepares to take off. Have the student take that mental picture with them and use it in the landing.

The takeoff is actually fraught with more potential for harm than the landing, because with each passing second the aircraft gathers more kinetic energy that must be dissipated in an accident. Safety lies with liftoff and escape from the hazards on the ground. In the landing, the reverse is true; each second that goes by puts the airplane in a less-hazardous energy state, so if the pilot manages to avoid disaster until near the end of the rollout there's little risk remaining.

Taking off is a transition, from a ground-bound machine taxiing at ever-faster speed into a living, breathing creature of the air. You can feel the wings gather lift, the wheels begin to lose their grip, and the controls take on the pressures of flight, trembling through the yoke and pedals. Ideally, as the transition proceeds, the airplane is placed in a takeoff attitude, nose held high and pointed at the departure end of the runway with rudder pressure, and the miracle of flight takes place by itself. Acceleration continues until the climbing attitude matches the engine's thrust, and the airplane stabilizes at V_Y, the pitch trim set to maintain a hands-off climb.

JUST PLAY IT BACKWARD

And so should the landing take place, simply in reverse order. As with the climb, there is a desired gliding speed, well above stall but not so fast as to prolong the flare into level flight that precedes touchdown. Power setting on the landing approach can be idle, or a minimal level needed to keep the glidepath on target, that is, staying out of the trees, but heading for the first third of the runway's length.

Imagine an open window at the end of the runway, through which the aircraft must be flown to effect a proper landing. The opening is about 50-feet wide and 50-feet high, and the airplane must be stabilized at a steady airspeed and sink rate as it flies through this window, lined up with the runway centerline. If this is done, all that remains is to level off into a nose-high touchdown and transition back into taxiing.

If, on the other hand, the approach is flown in an unstabilized manner, missing the window opening—or passing through it with one wing down, drifting sideways, airspeed off-target—a good landing is in doubt. Like a takeoff that gets out of control in its final stages, safety lies with escape aloft, where the runway edge lights are out of reach of the tires. Going around, as we practiced in the training for traffic patterns, is the best cure for a landing approach that misses the window at the threshold. Add full power, get the flaps retracted, pitch up to climb and fly straight ahead at V_Y, proceeding around the pattern for another try.

When abandoning a landing attempt, the student must consider what went wrong and do something to correct it on the next approach. If the airplane was too high to land, go out farther before turning onto base leg on the next attempt. If winding up over the grass along the side of the runway, hold alignment more diligently next time. If the plane ballooned in the flare, don't pull back so aggressively in the next landing. Don't do the same exact thing over again, unless it was a wind gust or other external influence that precipitated the go-around.

DON'T BE SHY

When nearing the runway threshold, it's important to keep the nose down to maintain airspeed until it's time to flare out into level flight for the touchdown. Intimidated by the sight of the runway suddenly rising up at them, students often exhibit "ground shyness" and unconsciously pull back as the runway nears, slowing into a semi-stalled state in level flight while the pavement is 25 feet below the tires, the entire width of the runway invisible behind a nose pointed at the sky. Without staring at the airspeed indicator, they must be taught to keep the nose down so the speed won't change while getting closer to the runway, then add just enough back pressure to level off at a height of one or two feet above the runway. In this position, they'll be able to see some of the runway on each side of the nose, determine that they're moving straight along the centerline, and take proper action to hold the airplane off until it touches down of its own accord, stripped of the energy needed to remain aloft.

What you're seeking, of course, is that elusive swish of tires gently brushing the surface with no jerk or squeal. To get that, several factors come into play. Airplanes in the air move not only up and down and from side to side, but also rotate out of straight-ahead alignment. To touch down smoothly the tires must be lined up with the direction of travel, which means using, or not using, rudder during the flare-out from the glide and in the holding-off period before touchdown. Sideward drift also creates a rough jerk as the aircraft's motion across the runway is arrested by the rubber ripping into the pavement. Drift generally comes from inducing aileron pressure when pulling back on the stick or yoke, or from uncorrected crosswind component.

THE GREASE JOB

For a soft landing, there must be no sink rate remaining, so the closer the student can get to the runway during the hold-off after the flare, the better the chances. One foot of space is good, six inches is better. But how are they to know? By having them remember their mental picture from takeoff and shifting their gaze to a spot about 100 feet down the runway, just left of the nose, while flaring out for the landing. They mustn't stare out the side window, or try to look through the engine cowling straight ahead. Show them that better depth perception is achieved by moving your eyes, not your head, to watch the runway that's visible to the side of the nose.

Sink rate can reoccur as a result of ballooning during the flare and hold-off, which brings the airplane back up to five or ten feet above the runway, the result of pulling back too much, too soon. Using the off-center shift of focus helps detect an early tendency to balloon, as does experience. The touchdown is best approached in stages, first flaring out from the glide to a nearly-level attitude, which is not suitable for landing, pausing for a second, then pulling the nose incrementally higher to reduce speed as much as possible. Soft landings come with low energy at contact, along with a lack of sink rate and side drift. To get there without ballooning in the flare, tell the student to plan on pausing for a second after flaring out, then adding a little more back pressure on the yoke, waiting for a tell-tale sink under the seat of their pants before adding back pressure to arrest that sink. Don't pull back further until the airplane tells you it's about to touch.

To correct a bad balloon, the pilot must stop pulling the yoke back until the ballooning subsides, then reapply back pressure as the airplane sinks to the runway, adding a touch of throttle to soften the sink. Then, take the power away as the tires hit and transition to taxiing. The landing is not over just because we're on the ground; keep flying the airplane while it becomes a land vehicle, holding the nosewheel off or lightly in contact and using ailerons to prevent leaning or drifting sideways. The nosewheel is not a landing gear, it's only a prop to keep the propeller out of the dirt. Touchdown should only take place on the maingear, with the nosewheel kept unloaded until needed for steering.

Crosswinds are just a fact of life in these days of single-runway airports. If the situation demands touching down on one wheel in a sideslip, holding up aileron into the wind and opposite rudder to keep the track straight, make sure the student does not relax the crossed controls just when they are needed most—at the moment of touchdown. Be ready to step in and keep the airplane out of trouble.

In the final analysis, we cannot make an airplane land, we can only put it into a position that permits it to land by itself. When energy is dissipated in level, straight flight, inches above the surface, the landing happens without further effort on the pilot's part. Our job is to put it in a landing attitude, just above the runway, and close the throttle; the airplane will do the rest. Always remember this motto: we will land no airplane before its time. To make the best landings, try to prevent the landing, until the airplane is ready to quit flying.

HEADS UP, CFI

Again, instructors need to sit up and pay attention when the ground gets close. We tend to get lackadaisical in the right seat when teaching familiar maneuvers out in the practice area, but there's very little time to react when a student goofs up at runway level. Guard the controls like a cobra, ready to strike. I've had many a student push instead of pull, trying to make the airplane land before its time. When the airplane darts for the edge of the runway, have your foot ready to stomp and a hand ready to slap a wing back down. Relax only when there's airspace to spare under the wings. Do not, however, "ride" the controls to interfere with learning.

If at all possible, I prefer to use full-stop and taxi-back patterns when teaching landings. The minute or so of ground time offers a quiet moment to critique what just happened, without the distraction of flying the airplane. The rollout and standing-start takeoff should be given equal status to the touchdown and liftoff. A touch-and-go, while sometimes necessary, is a hurried maneuver that skips many of the nuances needed for learning takeoffs and landings.

Landings are simple, when understood as the sum total of all inputs used during the approach, flare-out and hold-off. Stay close to the runway, stay straight with the centerline and reduce energy as much as possible. The airplane, then, does the landing by itself.

LESSON PLAN

OBJECTIVE

To give the student confidence in their ability to bring the aircraft safely back to earth.

TECHNIQUE

Isolate the individual components of the landing—approach, flare, hold-off and touchdown—and polish the skills needed for each. Demonstrate the correct attitude at each stage, and the patience needed while speed changes. Critique performance after each full-stop landing.

DESIRED RESULT

Students will correct errors on their own, achieving consistent safe landings and takeoffs during three or more full-stop patterns, on two successive days. Appropriate go-around decisions should be made without prompting.

CHAPTER 7

ALONE AND ASSURED

The first really significant achievement in one's life as a pilot is the initial solo flight. While it may not seem to be such a big deal in the grand scheme of things, in that the student is only repeating what they have done many times before, this time it's done without a CFI's safety net. There's no way anyone can get up there to help straighten out the mistakes. It's up to the student pilot to work out the solution and bring that airplane back to earth.

And so, the first solo really is a big deal. After soloing, the term *pilot* now applies in full appropriateness. Before, with the instructor sitting beside them, they were only a student. Now they have taken command and acted as the pilot of an aircraft, alone and unaided. Yes, the circumstances were controlled, and nothing was being done that hadn't been practiced many times before, but the fact remains—they did it all by themselves!

"When will I solo?" is an inevitable question asked by students. The only fair answer is, "when you're ready." To which I may add, "when I'm no longer needed." I never set a target date or training hour for solo. Everything has to align before a first solo flight can take place, and delays are routine. I know what I'm looking for, and I'm willing to wait.

What we're not looking for is perfection. The low-time student pilot is not going to exhibit a high degree of perfect performance. It'll be a long time yet before they will be private-pilot grade, and most licensed pilots still flub up occasionally. The goal is consistent and safe operation of the aircraft. Each trip around the pattern should be about like the ones that went before, with any mistakes corrected while they are still minor, so the result is a landing that's about the same as the previous ones. The instructor should have nothing to do with the process, other than to add some critiquing. Then, if the weather is benign and there's plenty of daylight and fuel remaining, the solo can take place.

THIS MAY NOT BE THE DAY

I've often considered that the next lesson will be the student's day to solo, and then subsequently run into a relapse that prevented me from turning them loose. If I see a good landing, followed by a sloppy one that should have been rectified before it got out of hand, and then one that exhibits another, unrelated error, it's time to turn this into a teaching period, curing the mistakes in hope that next hour will be better. Don't rush the Big Event just to keep on schedule. A student that takes 20 hours to reach the first solo will probably make just as good a pilot, perhaps with the same total hours, when they get to the checkride.

Sometimes it's just luck that determines who solos first. Because the wind picks up in the afternoon, the students on the p.m. schedule probably aren't going to get pushed out of the nest quite as soon as those flying in the calm hours. The student's readiness notwithstanding, a front may be threatening, or the traffic situation might be getting out of hand. No matter if commendable performance is being displayed, we can't take a chance if conditions aren't right.

What if the student screws up badly enough to require a go-around instead of making the landing? That's not necessarily a deal breaker. In fact, if the student made the decision to abort on their own, took the correct actions and then came back with appropriate alterations to prevent a repeat performance, I'll gladly trust that person with my airplane. Taking charge of a situation is the mark of a real pilot. On the other hand, if my assistance was required with the go-around, I'm probably going to give up plans to solo for that day, unless the rest of the hour goes flawlessly.

The instructor must restrain from verbosity when evaluating a student's readiness for solo. Do not prompt corrections and don't offer suggestions; wait for the student to make the first move, unless safety is an issue. I know we're paid to teach, but we also must see what the student will do on their own, when judging readiness for solo. You have to start dropping out of the loop at some point, instead of coaching and nudging the student through the traffic pattern. Hold your opinions until the circuit is finished, then offer your wisdom while they are taxiing back for another trip.

The decision to let the student solo probably will come after three successive, consistent, unprompted landings. Sign the paperwork without calling attention to your scribbling, preferably while the student is occupied with taxiing, and avoid long goodbyes, which only heighten the tension. I simply say, "you know, I haven't been doing anything for the last several times. There's no reason you can't do this on your own. Give me a landing by yourself; just keep on doing what you've been doing." And by then, I'm shutting the door and walking past the tail. Don't look back; let the student keep their mind on the job.

Yes, I have had students refuse to solo, out of a deeply-held fear or a dependence developed from too much dual instruction. My original mentor said, over a half-century ago, "at some point, they won't get any better until they solo," meaning you shouldn't wait until they are flawless pilots. Make sure your decision has been accepted, if no more than by silent acquiescence. Everyone has to agree, although most students will probably think they aren't ready unless you point out the obvious: that they've been doing it all.

Depending on the individual airport, I post myself in full view, so the student knows I'm watching, and so I can clearly see the airplane come and go. Or, if the control tower is situated conveniently, I may go up to the cab and watch from there. At an uncontrolled airport, I'll stay out by the runway, near the turn-off point. If the student's family members are on hand, I prefer to get out of the airplane while behind a hangar or far afield, so they won't know about the first solo until it's over. The student doesn't need the pressure of a crowd lined up by the runway, nor do I.

I don't feel it's necessary to have a hand-held radio clutched in my fist. If I expect to have to talk a student down, I shouldn't be sending them out for solo. When you turn them loose, it's up to them; you need to know you've done all you can do, and have faith in your efforts. That said, there should always be a means of communication available if unforeseen events arise, whether it's a portable unit, the Unicom or another airplane's radio. During first solos, I've had rogue traffic fly into the field against or without a pattern, I've seen winds shift, and once in a while a word of reassurance was in order after a couple of go-arounds were made.

IT'LL BE OKAY

You'll know how things are going by watching the takeoff; if the liftoff's straight and steady, you can expect the landing to be uneventful. Nervousness evident during takeoff will probably still be there at touchdown. All instructors can feel the airplane's motions when watching it in the pattern; we know the airspeed, hear the power being adjusted, and see the nose pitch down when flaps are extended. Usually, the first landing is the best of the traditional three trips around the pattern. I habitually flag down the student as they taxi past, walk up behind the wing and have a quick word and backslap before sending them out for another circuit. If the distance permits, I'll tell them to take it to the hangar, fuel pits or tiedown on their own after the last landing, while I walk respectfully behind. Another pilot has been born!

The standard post-solo speech reminds them that, momentous as it is, solo is merely the first rung on the ladder to a flying career. From now on, instruction is a shared responsibility; when flying solo, the student must be their own judge of when practice has gone well, or when another landing is in order. It's possible to get twice as much productive work done solo as when hauling an instructor around, because the student's full concentration is devoted to the task, rather than having to listen to the patter from the right seat.

I prefer to follow the Big Day with a dual review of everything that will be practiced solo, just to make sure all maneuvers are being correctly flown, and if conditions are suitable the next few hours can be solo periods. After three hours of solo practice, it's time for more dual; by then, the student is either bored and needing additional advanced work, or has accumulated frustration from not getting the maneuvers to come out right. Evaluation is in order, and more challenging work is introduced, until the student can do the full slate of maneuvers on their own.

Give the first solo its proper place in the time-honored tradition of pilot-making. It's a great first step on the way to the sky!

LESSON PLAN

OBJECTIVE

To allow the student to safely operate the aircraft alone in the traffic pattern, settling, once and for all, any doubt that they are a pilot.

TECHNIQUE

Solo is done under familiar circumstances, repeating a circuit of the airport that has been flown many times with the instructor. The first solo must occur under carefully-controlled circumstances, when the student's level of performance, the day's weather and all prerequisites are satisfactory. There should be no pressure to make it happen on a schedule.

DESIRED RESULT

Three circuits of the traffic pattern will be performed, in a manner identical to those flown under the tutelage of the instructor. Once accomplished, further training will become a shared responsibility of the student and CFI.

CHAPTER 8

PUSHING ON

Once past the major milestone of an initial solo flight, a student pilot is qualified to practice briefed maneuvers on their own, under the supervision of a CFI. Supervision means the instructor must know when and where the solo flight is taking place, that the weather and daylight is satisfactory for the individual's capabilities, and what the day's practice session will contain.

I have never felt this can be done from a long distance; I want to be on-site for the preflight briefing, departure and recovery, at least until the student has proven to be trustworthy. Over time, I may simultaneously conduct a local dual flight or take care of administrative chores in the area. However, I never authorize solo flights while I'm out of town. Instead, other instructors I trust will perform this duty for me, in exchange for similar backup duty for them.

Once all the pre-solo maneuvers have been reviewed, I typically expect three hours of solo practice to be flown, gradually reinforcing the student's confidence in their ability to fly out to the practice area and return, performing the full repertoire on their own. These solo hours are important steps in a pilot's development; too much dual time encourages dependency and dulls decision-making. Even though the student is being supervised from the standpoint of a CFI dispatch, minute-by-minute choices have to be made by the pilot-in-command sitting in the left seat. Not only is solo time more productive in terms of allowing uninterrupted focus on the task, it also increases vital self-confidence. No student should be sent out solo unless there have been discussions about what to do if unforecasted weather moves in, or if they become disoriented in the practice area.

RULES OF THE GAME

I've always used two designated practice areas, on opposite ends of the prevailing wind axis, with solid geographic boundaries that can followed back to the airport. This allows a student to depart and climb out while heading into the day's wind, practice their routine upwind of the field, and come home quickly with a tailwind for an easy traffic pattern entry. If practicing downwind from the airport, recovery at the end of the session, or coping with a crisis that requires a quick return, can take nearly twice as long. I try to limit solo traffic pattern practice to three takeoffs and landings, done in the last half of the scheduled hour after engine temperatures have stabilized. This keeps the student from spending an inordinate amount of time on what should be only a small portion of their training, taking off and landing. And it should be made clear that there's not supposed to be any solo practice of forced landings, or with the instrument-flying hood in place. Don't laugh, it's happened.

Once the initial three hours of solo are logged, a dual re-evaluation is in order. Some students will have goofed off for the three hours, showing little improvement in the basic maneuvers they were assigned to practice. That time wasn't totally wasted, from a pilot development standpoint, but they need to have it pointed out that time spent circling a girlfriend's house doesn't get them a license. Normally, there should be a detectable increase in proficiency when the dual review is done, as they've polished their flying while doing solo stalls, slow-flight, S-turns across a road and traffic patterns. At the same time, while checking progress we should also introduce some advanced maneuvers, to keep up the pace of training.

SHARPER TURNS

If the student exhibits mastery of 360-degree medium-bank turns, it's time to move on to steep-bank turns. The basic requirements of maintaining control in a bank greater than 30 degrees have already been taught, in the course of learning to maintain a ground reference track around a turn point with a tailwind. However, we are now expecting the student to keep precise altitude, airspeed and orientation during a sustained steep-bank turn at a higher altitude, rolling out exactly on the entry heading. To teach the maneuver, I prefer to continue through 720 degrees, two complete circles, because the rate of turn is so rapid that a single 360 is over almost before it begins. In addition, the second orbit will usually require some control inputs to deal with the disturbed air left by the first turn, good training for a rough-day checkride.

The target bank angle is 45 degrees, steep enough to exhibit the over-banking tendency that must be countered, and also sufficient to require noticeable extra back-pressure to hold altitude. But, it's not so steep as to degrade a training airplane's performance and generate high g-loads. As the bank increases beyond 30 degrees, it's necessary to add and hold extra force on the control yoke, to keep the nose from drooping below level-flight attitude, and to advance the throttle to at least maintain entry RPM with a fixed-pitch propeller, so airspeed won't sag under the g-load. As with slow-flight and stalls, a minimum of 1,500 feet of altitude above ground level is to be held, to assure a safe recovery if control is lost.

Many students timidly reach the 45-degree bank only after turning 90 degrees or more, particularly when rolling to the left, where they are more likely to perceive a risk of falling out of their seat. The airplane should be locked into a 45-degree bank by the time it turns through 30 or 45 degrees of heading change. The rate of roll, both during entry and recovery, should be continuous and moderate, blended with the required shift in back-pressure needed to balance the changing g-load. Another common error is staring at the altimeter needle, rather than using the horizon as a pitch reference, resulting in a roller-coaster ride around the turn, "chasing the needle." A steady steep turn requires quick correction of any dip of the nose from level flight. Otherwise, speed will build up, perhaps exceeding maneuvering speed or the placarded maximum for steep turns, and a heavy g-load will be required for recovery unless bank is reduced.

When correcting small losses of altitude, most students will pull back, feel the g-load, and immediately relax control to avoid the unsettling sensation of seat pressure. Instead, that g-load must be sustained long enough to halt the descent and bring the nose back to the horizon. Ideally, the deviation never occurs, because a constant mild g-load was held from the moment the airplane stabilized in a 45-degree bank. One should never feel light in the seat during a steep turn; that's an indication that you're about to lose altitude. First the nose goes down, then the rest of the airplane follows; act quickly, and the altimeter needle will stay steady. To demonstrate, I twist in a couple of turns of elevator trim as I roll the airplane past 30 degrees of bank, pointedly releasing the controls in the steep bank to show that steady g-load means a constant altitude. A nudge of opposite rudder to offset the overbanking tendency goes unnoticed.

DOWN ON THE DECK

Once the student's performance in slow flight and stalls has been reviewed, it's a good time to initiate a forced-landing emergency, which hasn't been experienced since the last dual session. This consumes some no-longer-needed altitude while establishing a glide, choosing a landing site, running the restart checklist and setting up the approach. After the simulated forced landing is abandoned, the remaining altitude will be convenient for reviewing ground reference maneuvers.

If not taught previously, I will introduce turns around a point, an extension of S-turns across a road and rectangular courses which require on-going concentration to maintain a constant radius of turn and altitude, rather than allowing the student to make a fresh start each time a road is crossed. Prominent road intersections, section lines or other bisecting landmarks are better targets than a single tree or pond. The student must understand that the goal is not to keep the wingtip pointed at the chosen spot, but rather to offset wind drift, keeping a constant distance from the point by varying bank angle and crabbing while flying around the circle. Precise altitude control requires adjusting back pressure for the varying bank angles on the upwind and downwind sides of the circle. Entry on a downwind heading produces an easier-to-control start to the maneuver.

PATTERNS WITH A PURPOSE

To further evaluate and upgrade performance, we will return to the traffic pattern and check the student's progress in takeoffs and landings. If they are up to speed, showing a normal grasp of pattern work, it's time to move on to more challenging short-field and soft-field takeoffs and landings. You don't need to be an instructor very long to understand that no two students are the same. Adding precise steep turns and turns about a point during an hour of review may be enough for some students, requiring a couple of hours of dedicated solo practice before introducing the specialized takeoff and landing tasks. To avoid confusion, I sometimes show the short-field operations in one lesson and soft-field technique in another. As long the student is able to absorb it, however, we need to continually add more tasks to the repertoire.

GETTING UP AND OVER QUICKLY

Short-field takeoffs are designed to clear an obstacle in the takeoff path by using the maximum performance available from the airplane. Instead of rolling out onto the runway and smoothly adding power while on the move, the airplane is positioned at the extreme end of the pavement with the recommended flap setting, brakes are applied, full power is brought up, and the binders are then released in order to gain every foot of advantage. Because the climb speed target at the barrier is V_X (best-angle-of-climb speed), the liftoff must be initiated well before V_X so that the wheels can leave the ground without a sudden yank on the yoke. The nose is raised to slightly higher than normal takeoff attitude, timing the acceleration after breaking ground so as to reach V_X just as the obstacle is cleared. Once a stable V_X climb is achieved and no further obstacles need to be avoided, the nose is lowered to increase speed to V_Y and flaps are retracted.

Short-field landings are merely precisely-flown versions of a normal landing, designed to arrive on or just beyond a designated target spot. To do this, the approach is made with maximum drag and minimum airspeed, keeping power available to control the glidepath. I like to see the landing begun with a power setting that will carry the airplane to the runway with progressive flap extension and speed reduction, requiring just incremental reductions of power or no power adjustment. Moving the throttle back and forth, barring updrafts and downdrafts, shows a lack of planning.

In most cases, the final approach speed for a short-field landing is 1.3-times the stall speed in landing configuration, adjusted for static position error, plus a few knots for a student safety factor. The secret of touching down on the target zone is timing the reduction of power to idle. Cut the power too early and you'll touch down short; leave it on too long and the float will be excessive. The airplane should decelerate in level flight after flaring out, reaching a normal touchdown attitude just as the desired spot is reached. No attempt should be made to dump the aircraft onto the runway nosewheel first, or to try to reach a full stall. Braking technique should follow the POH recommendation, bringing flaps up if needed, holding the stick back, and using firm braking without sliding tires.

A SOFT TOUCH

If the student is ready for the added challenge of soft-field operation, some time must be spent briefing the procedure to make sure they understand the difference in objective from the short-field takeoff and landing. Yes, there can be a soft *and* short field, but it's better to teach two distinct techniques, which can easily be combined out there in the real world. Both the liftoff and touchdown need to be made at the slowest possible airspeed, and since this means operating right at the stalling speed, soft-field work must be done with care.

The soft-field takeoff is begun with flaps positioned for maximum lift, moving steadily onto the runway to avoid stopping and getting stuck in the simulated mud. Once power is applied, the nose is brought up to position the wing to generate high lift and to remove the nosewheel from the drag of the surface. This means there's no nosewheel steering and forward visibility may be restricted, so vigorous rudder use and peering around the elevated instrument panel will be needed to stay on the centerline. It's possible to over-rotate and drag the tail in some planes, so the CFI must be alert to avoid damage. The airplane will take off by itself as soon as it reaches minimum unstick speed, at which point it's necessary to relax back pressure to level off within 10 feet or so, maximizing acceleration in ground effect. If no obstacles are ahead, flaps are brought up and V_Y is established.

A soft field landing assumes there will be rapid deceleration without the need for braking after touchdown, because the surface is draggy, thus a larger touchdown target zone is acceptable. As with the short-field approach, full flaps are extended to achieve a lower stall speed and power is used to control the glidepath. I prefer to add a few knots of approach speed for improved control, since stopping won't be a problem. Full flaps are extended on final, the flareout is made a bit higher than usual to avoid a premature touchdown and, if desired, some power is retained through the touchdown to achieve a slower landing. The stick should be all the way back, against the stop, just as the tires touch, and kept there during the rollout. Students frequently confuse a soft-field landing with a whisper-soft touchdown. The idea is to fully stall the airplane to minimize the stress of a rough or soft surface on the landing gear and airframe, which doesn't necessarily mean the landing itself has to be a grease job. However, removing all excess airspeed does generally create a softer landing.

If these additional maneuvers are written down correctly in the student's notebook (the instructor should review all notes taken), they can provide enough diversion for an additional three hours of solo flying, at which time we'll have another dual session to make sure proper progress is being made and no errors have crept into the technique. Pushing on to advanced work is really about learning how to use the airplane's maximum potential.

LESSON PLAN

OBJECTIVE

To advance the student's skills by dual review, solo practice, and introduction of progressively more-demanding maneuvers, based on those already learned.

TECHNIQUE

Tailor the amount of new material introduced to the ability of the student, making sure fundamentals are firmly in place before advancing to the next level. Have the student add written step-by-step instructions in a personal notebook, for review during solo sessions.

DESIRED RESULT

By the end of the dual/solo practice phase leading up to cross-country training, the student will be familiar with all maneuvers that will be required on the practical test, and be capable of practicing them solo. Confidence will be increased by an increasing amount of solo pilot-in-command flying time.

CHAPTER 9

THE WORLD BEYOND

Even for pilots espousing a desire to remain close to home base, flying just for fun, there will be a need to stretch the wings and fly over the horizon. Airplanes are built for travel, as much as for release from the cares of an earth-bound existence. I've flown a Kitfox on a 200-mile hop and enjoyed the ride just as much as making a 500-mile trip in a Bonanza or flying halfway across the continent in a Skylane. Cross-country flying is the epitome of lightplane utility, and it's the goal of the next phase of our training.

The portion of the curriculum completed thus far has concentrated on acquiring the skills needed to maintain control of the aircraft. The student has learned to avoid the hazards of flying too slowly, to avert the wind's opposing force while keeping a ground track, and to maximize the airplane's performance to accomplish specific tasks. Now, it's time to lay this accumulated wisdom aside, or at least keep it on call in the background, while we learn to hold a long, straight point-to-point course.

That's the reason we've been practicing so hard to perfect the maneuvers learned earlier. When we're engrossed in the cross-country phase of training, it's not unusual to see some atrophying of stick-and-rudder skills, so they will have to be sufficiently ingrained as to be readily at hand when called upon. Wise students, and their instructors, will spend an hour in diligent review practice once in a while, perhaps when weather conditions aren't suitable for the planned cross-country trip.

LET'S MAKE A PLAN

Planning is the basis for successful cross-country flying. Any trip begun with only a sketchy sense of where we're going and how to get there is bound to end with doubt and uncertainty at best. At worst, you'll land somewhere lost and frightened. Plan your flight and fly your plan. Yes, you can be flexible, ready to deal with changes dictated by weather or unforeseen circumstances, but such flexibility is part of your planning, simply diverting to alternative routes or airports that were pre-selected.

The basics of flight planning are to know how far you're going and how you're going to make it happen. Divining the distance to be covered allows us to roughly determine the time and fuel required. Measuring the compass course sets the stage for selecting a route that avoids hazards and workload.

Weather acquisition is critical, even if it's a clear day, because you can't be sure what's over the horizon or moving in later. Getting the weather is not as simple as it once was, in the days of a friendly neighborhood flight service station on every block, so students must know how to gain access to weather, along with TFRs (temporary flight restrictions), NOTAMs (notices to airmen) and SUA (special use airspace) info.

Beyond this bare framework, it's also important to acquire details needed to use the airports, air traffic control and navigation facilities along the way. It's much easier to look up a frequency that's been written down on a pre-planned flight log than to find it on a sectional chart, often buried in a jumble of terrain and other symbology. A few reminder notes about upcoming high ground, tall towers or airspace boundaries add to the safety of the trip.

Teach students to be organized. They need to write all this stuff down in an orderly fashion, so they can readily refer to it a couple of weeks later, when they're finally ready to make the trip they've planned. It's more important to have the plan in a format that's easy to interpret than to meticulously fill in blocks of numbers, so devise a simplified flight log form if you wish; the goal is just to be able to find that tower frequency when it's needed.

We will be teaching the basics of aerial navigation, even if the airplane is outfitted with the latest GPS moving-map navigator that instantly tells us the direction, distance, time and speed to our destination. We never want to be so dependent on technology that we can't do without it. No piece of equipment should be so valuable as to jeopardize the safety of the flight if it fails. Therefore, we'll begin by discussing three independent methods of navigation.

First, there's pilotage, which simply means map-reading. One looks at a map, compares it to the view out the window, and determines where one is by the landmarks seen, picking another spot in the distance that leads to the destination and heading for it, thereby leap-frogging across the countryside. Following a string of towns along a highway, railroad or river is basic pilotage.

Next, we will use dead reckoning, a term corrupted from "deduced reckoning," to determine our position. We can hold a compass direction for a specified length of time and, knowing our speed, determine where we've gone by computing the distance covered, using a manual E6-B or electronic flight calculator. No landmarks are involved, merely careful attention to the math and accuracy of the inputs; Lindbergh successfully flew across the Atlantic by dead reckoning, so it works.

Finally, we can use the electronic navigation equipment in the aircraft, both as a means of holding a course and to verify position. The VOR system is easy to use and doesn't depend on a delicate signal from vulnerable orbiting satellites. Today's modern GPS boxes can do much more than just generate a pink line to follow on the display, but all that capability comes at a price in complexity. Whatever electronics are installed, we may have three systems of navigation working at the same time; pilotage, dead reckoning and radio.

It's very important to always have at least two methods of navigation working continually, so one procedure can verify the other's findings. Each system is subject to human error or failure, so by cross-checking one with another we'll spot any discrepancy before it leads us astray. With pilotage alone, it's easy to hastily misread a map, making an assumption that the town we're passing is the one we're looking for. Running a simultaneous dead-reckoning computation will tip us off to the fact that it can't be *that* town, because the current time doesn't say we could be there. Conversely, if we misread our watch, mis-calculate the compass heading or forget to reset the directional gyro, the landmarks on the chart won't match our prediction, showing that the dead-reckoning results are dead-wrong. Radio fixes can settle conflicts with the map if landmarks are scarce, and pilotage will serve to verify the correct electronic course. So, teach students to never rely on a single source of navigation, but to always verify the navigation with a second system.

LAYING OUT THE TRIP

In its most basic form, cross-country flight planning involves drawing a line on the chart from the departure airport to the destination and following it. This is exactly what the GPS navigator box is doing. However, one always has to consider this preliminary step in terms of what lies between. A direct route is fine over open, flat country, but may need to be altered to avoid high terrain, open water or unfriendly large cities. An easier route may only involve flying a few miles further.

I prefer using a bright red pencil line to mark my course, rather than an ink pen or a marker, an erasable medium that allows me to correct a mistake and a bright red color that stands out from the other tints on the map. At night, of course, a black pencil is used because of the red cockpit lighting. Once the courseline is laid down, we measure the overall mileage and enter it on the flight log form, then apply the plotter's protractor to a true-north grid line to determine the TC (true course). This course is the starting point for the calculation of wind-drift correction and the subsequent adjustments for magnetic variation and compass deviation, each duly noted on the log form.

Once the how-far and which-way numbers are determined, it's time to pick out en route checkpoints for pilotage and short-term dead reckoning purposes. The first checkpoint needs to be within easy visual range of the departure airport, so it can be seen immediately after leaving the traffic pattern. Many pilots have gotten themselves lost over unfamiliar country because they couldn't locate the courseline and had already flown out of sight of the takeoff field. Keeping that first checkpoint within five to ten miles, where the airport can be seen behind the airplane and the checkpoint is just ahead, makes it easy to line up on course. Ideally, we will reach cruise altitude and accelerate into level flight about the time we're over the first checkpoint.

Succeeding checkpoints should be about 10 minutes of flying time apart; closer together needlessly adds to the workload, farther apart invites wandering off course. This equates to about 15-20 miles at trainer speed. Suitable landmarks should be easily visible, yet precise; a large lake is clearly seen, but there needs to be a cove or point of land to define the crossing point. Prominent road patterns help designate a ubiquitous small town. Bridges accompany streams, railroads and freeways, and will have bridges or overpasses where they cross. Powerline and pipeline rights-of-way break up the unbroken expanse of timber country. Poor checkpoints include small skinny towers, powerlines in prairie country, villages without unique features and distant towns too far off course to help determine position.

As the student enters each checkpoint on the flight log, we will note the elapsed mileage from the last checkpoint in an adjoining column, with a line separating it from a "miles to go" number, a declining balance that shrinks as the airplane passes each checkpoint. In that way, we can always have a distance with which to compute an ETA (estimated time of arrival) at the destination once an actual ground speed has been established. When finishing the laying out of the trip, the leg from the last checkpoint to the destination should equal the remaining "to go" number within a mile or two. If it doesn't, a measuring error was made or the math is wrong. This double-check is important, so we will catch such a mistake before we depend on the flight log during the journey.

Adjacent to the mileages column, we'll leave room to note two time numbers as each checkpoint is passed during the trip, an estimated time when we expect to reach the checkpoint and the actual time of passage. Only the minutes need to be written down; we'll undoubtedly be aware of the hour when the aircraft took off, so there's no need to write four numbers when two will suffice. There's no point in writing down the elapsed time expected to be flown between checkpoints; once it's been computed, we'll simply add it to the time-over from the previous checkpoint and write down the minutes after the hour when we expect to arrive at the next checkpoint.

A "remarks" column is left on the margin to give space for notes, such as a new heading to be taken up at a checkpoint, a hazard or special-use airspace we don't want to ignore, and reminders to check in with ATC (air traffic control). I like to write down the arrival frequencies in their order of use; begin with a local FSS (flight service station) outlet, then the ATIS or AWOS for weather, followed by the correct approach control frequency if needed, then the control tower and ground control frequencies. It's much handier to have these written down on the flight log in large format instead of having to keep finding them on the cluttered chart.

We now have the flight planned, or at least one leg of it. Using the flight log, the trip should go smoothly, as the student keeps track of their progress using at least two, or perhaps three, navigation methods at all times. The workload is actually lessened by having the trip well planned; about one minute of notation and calculation at each checkpoint will keep up the log, leaving the rest of the time to enjoy the scenery.

LESSON PLAN

OBJECTIVE

To learn how to utilize the necessary navigation tools and materials to perform a future cross-country flight, assuring success through proper flight planning.

TECHNIQUE

Use a sectional aeronautical chart to lay out a three-leg trip, with each leg encompassing 45 to 90 minutes of flight time. Emphasize that paper charts are not battery dependant and a flight log form should contain all information relevant to the trip. Plan on using redundant navigation methods throughout.

DESIRED RESULT

After demonstration by the instructor, the student should be able to lay out additional legs of a trip, measuring distance and course, choosing checkpoints and noting relevant data. With planning completed, the trip is ready to be flown.

CHAPTER 10

THE PROVING RUN

The purpose of flying a dual cross-country is to make sure the knowledge acquired in ground school and in laying out a trip can be put to work in the airplane, which is admittedly one of the worst classrooms in which to teach. However, while it may be cramped and noisy, there's no substitute for the real thing, with real scenery passing by outside.

Several skills are to be developed during the course of the dual cross-country flight, which should actually be a triangle of three flights flown in sequence, providing a break every hour or two and allowing different country to be seen throughout the trip. To recap the previous chapter, the obvious first skill to be learned is navigation, a consolidation of pilotage, dead reckoning and radio usage. Second in importance is the integration of ATC services into cross-country travel, which can include contacts with flight service for enroute weather updates and using radar flight following on the final leg when most of the dual instruction is past.

Another set of critical skills to be acquired are the techniques applied if one becomes lost, or "disoriented" in the polite vernacular. Also important is the ability to take care of the aircraft when away from home; how to park the airplane, tie it down, obtain fuel and secure it against intruders. These are all things that are new to a local-flying student pilot, but they will soon become their responsibility.

Make sure the trip will visit airports that are different from the home field; if the student is based at a controlled airport, choose one or more busy uncontrolled fields, where all decisions are up to the pilot, using information obtained from Unicom and other aircraft. On the other hand, if they have only flown from a non-tower field, don't waste the dual cross-country on similar airports; go to the biggest, baddest controlled airports you can find. You're being paid for training, so look for teachable moments, not repetition.

Don't launch until the trip is planned and briefed in detail. Don't try to make up for inadequate preparation in the air; that never works out well. Fly a route as planned and it'll go off like clockwork. You may want to take the opportunity to file a flight plan with flight service, just for added radio practice and for future use by the new pilot. VFR flight following is a good substitute for a filed flight plan, but the chatter of ATC will interfere with instruction initially, so it may be best to defer it until later in the circuit.

THINGS TO AVOID

Many flight schools make the first dual cross-country a super-easy route, by following a major highway to a simple, easy-to-find airport, out and back. They even have the student repeat the trip solo the next day. What's been taught by this exercise? Very little, in terms of a student's ability to navigate on their own after they get their license. I feel it's better to strike out on a course across open countryside, or perhaps along an airway route, and to create a triangular circuit that'll involve different types of airports. And the subsequent solo trips should be made to entirely different airports, an accomplishment that gives students confidence that they can make a foreign trip by themselves. Naturally, the routes and airports approved for solo cross-country should only be ones that have been visited and verified by the CFI as suitable for students.

Never push your luck on the weather required for a student cross-country. Realistic training under less-than-ideal conditions can offer valuable teaching opportunities, but so can cancellations made after a thorough examination of the weather data. Students need to know there are times when even the most seasoned pilots don't go. An interrupted trip results in uncompleted training, which must be repeated at added expense, so we'll go only when there's a reasonable chance of success. Pushing on through risky weather gives a student exactly the wrong message; make sure they see you respect limitations that are higher than just the legal minimums. And make it a point to get an early start to beat the turbulence. Students have enough trouble learning to navigate and hold a course without the pounding of rough air.

As you depart on the first leg, the student may be overwhelmed by the sudden workload, requiring you to assist in sorting it out. Show them the priorities; aviate, navigate, then communicate. "Aviating" refers to simply flying the airplane on a straight course; keep airspeed or altitude within reasonable limits, stay on the heading by making continual corrections and take care of the power management. At the same time, it's important to navigate, by staying aware of one's position; no one ever got lost by keeping track of where they were. That means just noticing a landmark sliding by, noting the time if applicable, and remaining cognizant of the passage of time. If it's been five minutes since you last confirmed your position, you're obviously no longer there; teach the student to run their finger along the course that's been made good on the chart and see if something recognizable isn't about to turn up in the windshield. Yes, a GPS line to an active waypoint is a great tool, but even the best panel-mounted map is a sketchy substitute for the view out the window. Verify the GPS information.

Communication may sometimes have to be pushed to the top of the duty stack, but don't be deceived; we always need to be flying the airplane first, and to continue navigating in the background even when talking. It only takes a few seconds to respond to a radio call, then one can go right back to the cockpit duties. Point out the importance of being organized early when heading toward a full-service airport; have the frequencies jotted down in order, think about what you're going to say, and listen to the radio traffic for a while to find out what you can expect to hear. Role playing, with the instructor taking the part of the controller, should be done both during the preflight briefing and in flight, just before pushing the mic button.

Particularly on the hectic first leg, allow the student a bit of tolerance while they are learning to manage the three priorities. If they get off altitude by 200 feet, point out that it should be corrected when there's time, but they shouldn't obsess over busy-work at the expense of other, more pressing duties. As management skills are acquired, the tolerance can be tightened up.

Getting lined up on course initially is critical. In the early days it was common practice to climb in a circle around the departure field before leveling off to start a bee-line toward the destination, across the airport. That may be frowned upon today, but we should be looking for a starting gate in close proximity to the airport and check the compass and directional gyro as it is passed. We don't want to lose sight of the departure airport until everything looks right.

During the course of the three-legged dual cross-country, the CFI must gradually loosen their grip on the situation, allowing the student to stumble through full responsibility for the flight by the end of the third leg. Never mistake willing acquiescence for confirmation; test for knowledge by asking "what if" questions and by observing the student take charge without prompting. To be qualified for a solo cross-country signoff, the student must find their way home without help on the last leg and keep up a proper log of times over checkpoints, plus be capable of handling the communication duties.

GETTING UNLOST

To teach the techniques of surviving disorientation, I often ask the student to don the instrument hood for some practice, saying, "I'll take care of the navigation for a while." Then I'll have them turn to headings that take them off course a few miles, climb or descend and generally wander around for a few minutes. After this, I'll have them take the hood off on a heading that's a bit far from the one required to take them home and I'll announce, "I'm lost, find out where we are." The logical steps are to turn back to the desired heading, look outside for something that might be on the map, and perform a VOR bearing check to confirm position, or use the GPS's nearest-airport or direct-to features. Emphasize that it's no sin to be lost, but it's vitally necessary not to stay lost.

If there's a lack of ability displayed, more training is obviously needed, but we'll not repeat the same trip. A more-focused review will be undertaken the next time, emphasizing whatever was found to be unsatisfactory on the first expedition; holding a course, computing groundspeed from times over checkpoints, communicating with ATC or staying ahead of the airplane. Sign off no student until they are ready to make a trip on their own.

LESSON PLAN

OBJECTIVE

To transfer the flight planning and ground instruction to actual flying of the planned trip, resulting in bilateral confidence in the student's ability to perform solo cross countries.

TECHNIQUE

Use the trip log, charts, flight computer and onboard equipment to keep the plane on its course, verified by multiple navigation techniques. A phased-in transfer of piloting duties progresses as the various legs are flown, until the student is performing unaided.

DESIRED RESULT

The student arrives at home base after acquiring and maintaining the homebound course, keeping track of position and writing down times over checkpoints. They should feel confident of their ability to fly a simple solo cross-country.

CHAPTER 11

BABY STEPS, BIGGER STEPS

Solo cross-country flying needs to be approached carefully, with due regard for the student's fragile confidence and the risks involved. After the lengthy dual cross-country training, they should be well grounded in the techniques involved. However, the final proof is going to be in the student's ability to return from a solo voyage beyond the confines of the practice area.

The first solo cross-country should be stone-simple, yet meet all the requirements for loggable cross-country flying time and a practical demonstration; it must include a landing more than 50 nautical miles from home and should visit two airports, each separated by several good checkpoints. I nearly always use the same easy triangle for all my students, a route that's bisected by a major highway leading back to home base. The airports involved are similar in size to the student's point of origin, radio requirements are rudimentary and total flying time required is no more than two hours. I prefer to use different airports for the solo cross-country than were visited on the dual training trip; doing so gives students confidence in their ability to adapt to strange surroundings. However, careful briefing is done.

AWAY ON THEIR OWN

Weather conditions should be totally benign for this first trip. Winds aloft should be light enough that the forecast can be off by several degrees or knots without jeopardizing the outcome, and no ceiling or visibility problems should be threatening. I will not allow the student to launch unless there's well over an hour of daylight remaining after the most pessimistic imaginable flying time required. Some students drag their feet interminably while folding and refolding charts and giving the ship one more lengthy preflight. Know their habits and plan accordingly; set a deadline time for the takeoff, beyond which cancellation is in order.

I'll review the student's planning carefully and pose some what-if scenarios based on the day's challenges; what if the wind picks up and you can't safely land? What if you don't find the airport? What if the radio fails? All the answers should basically be "return to base," using the major highway as a pilotage-aid if needed. I will then endorse the student's logbook for the trip, sometimes adding authorization for an alternate field.

Although the trip is going to be well-monitored and tracked, we may also file a flight plan, just for familiarization purposes, but due to the short leg lengths involved we'll make it a round-robin plan, involving only starting and stopping contacts, rather than opening and closing multiple flight plans at each stop. A radio call to flight service will activate the flight plan on file, and if the student neglects to close it by radio when in sight of home base it can always be cancelled by telephone.

The drill is to launch on course, using the appropriate direction of departure for the day's winds, find the vital first checkpoint, keep a log of times and groundspeed and make a full-stop landing at two unfamiliar, non-challenging airports. A time-honored ritual involves presenting one's endorsed logbook for a signature in the "remarks" column; anyone can perform this duty, usually the lineperson who waves the proud student into a parking slot, or an interesting-looking customer service rep. After a quick potty and pop break, it'll be time to taxi out for the next stop.

After collecting the two signatures (no penalty is assessed for neglecting this duty, if there's no one around), the student heads for home. The final leg might only be a half-hour long, but there must be two hours of fuel on board for the departure or a refueling is required. Yes, I send the company credit card along with the airplane; the risk of fraud is certainly no greater than entrusting a valuable aircraft to a low-time pilot.

I do not normally ask the student to call in to home base at each stop, unless there's a question needing an answer or some other difficulty. They are responsible for the actual conduct of the flight, with the instructor's aid always available, so no news is expected to be good news. Each of the airports used has been visited personally by the CFI involved, making sure the facilities are easy to use and friendly for students.

There's nothing wrong with prefixing all radio calls with "student pilot" added to the aircraft ID, giving Unicom or tower operators a heads-up, in case they don't recognize the tense delivery of a beginner. The first solo cross-country is always directed toward low-pressure communication requirements, but after a few trips the student realizes those other voices are just coming from normal people.

Occasionally, a student can suffer a relapse and, despite all the training and planning, fails to find one or both airports. A post-flight critique will usually uncover the reason; the log of checkpoints shows no entries for the time of passage, the student has no memory of seeing the associated town, and a recheck of the directional gyro and compass alignment probably wasn't done. A supplemental dual review to brush up on the basics of pilotage and dead reckoning should be scheduled right away; do not dispatch the student solo again if there's any doubt of their ability.

LET'S DO IT AGAIN

Assuming the initial solo cross-country is successful, the bar is raised for the second trip, directing the student toward more distant points and slightly busier Class D airports, logging perhaps three hours of flying time overall. None of the trips, however, should cross challenging terrain or vacant expanses. Students should see well-designated landmarks under their wings, not unbroken timber. Every lake looks alike when you are just beginning your navigation career.

Assuming trip lengths of an hour or so, refueling is normally necessary at one or both of the more-distant points. The procedure of self-fueling should be part of all cross-country training, even though we normally expect to use airports that offer line service. It's advisable to call each of the FBO's in advance, to make sure that fuel is available and that hours of operation haven't changed.

I prefer to have three solo cross-country trips in the student's logbook, each one increasing in difficulty. One or perhaps two can be lucked out; three successful trips will prove to everyone that this pilot is indeed qualified to be a private pilot. At one time, ten hours of solo cross-country flying were required to qualify for the license, a not-unreasonable amount, but the requirement was somewhat arbitrary and sometimes presented difficulties when a student had caught good winds on all three of their trips, resulting in a logbook that was a few tenths short of the requirement. Current regulations allow applicants to qualify with only five hours of solo cross-country time, which really isn't enough.

ONE MORE TIME

The final three-legged flight will be to some reasonably-busy Class D or higher airports, perhaps involving a contact with approach control to reinforce what was learned on the dual cross-country. Quite honestly, every student is different, and there are some who can't quite rise to the challenge of the busier airports, so if I detect a hesitancy on their part I may assign them to a less-challenging finishing-up trip, until they can gain more experience. Do not attempt to stamp every pilot out of the same mold; not every person learning to fly is headed for an airline cockpit, nor do they all plan to fly major expeditions to distant lands. Our job is give them the basic training and experience to be a private pilot, from which they can venture forth to pursue their personal goals.

By the time all three solo cross-country triangles have been flown, the student knows they can make a trip without difficulty. No doubt some weather, turbulence and wind will have been conquered, and the radio is no longer emitting foreign tongues. It's now time to get back to precision flying, polishing the skills learned earlier.

LESSON PLAN

OBJECTIVE

To instill confidence in the student's mind by flying three triangular solo cross-country trips, visiting airports never seen before. Meeting the minimum requirements for the private pilot rating is of secondary importance.

TECHNIQUE

Careful flight planning and cautious approaches to weather and daylight will assure successful, confidence-building outcomes for each flight. By stepping up from a very simple and short initial solo cross-country to progressively longer and challenging trips, the student gradually acquires added skill and assurance.

DESIRED RESULT

Students will finish all three solo cross-country flights without difficulty and be qualified for using a private pilot license once it is obtained. Leaving the home airport for a distant destination should hold no trepidation.

CHAPTER 12

IN PURSUIT OF KNOWLEDGE

The private pilot knowledge test, or written exam, has evolved over the years. As one of the prime prerequisites for the checkride, passing the written has proven to be a very tall hurdle for many student pilots. One of the failings of many flight instructors is a tendency to concentrate on the flight training, ignoring or diminishing any emphasis on academics.

This is unfair and failure-inducing. There's much to learn about airplanes and the way they are to be flown, and there's obviously not enough time to go over it one-on-one, passing information across a classroom desk. Instructors must assign homework, then verify that it was done, and they must continue to give additional assignments, all designed to integrate with the syllabus.

It's not all that difficult to check on a student's progress with the academics. In any discussion about the preflight, or the latest maneuvers, or the day's weather and environment, a few questions will determine who's been looking up the facts, and who hasn't. One of my students recently inquired, in the course of completing the runup checklist, "what happens if there's no reading on the suction gauge?" Which rather shocked me, because he had 50 hours and was in the final phase of his preparation for the checkride. He should have read about or been shown the flight instruments that depend on the vacuum system long before now. Obviously, I had failed to adequately verify that he'd done his homework.

THE WRITTEN

Sitting for the formal knowledge exam has become a big deal from the days when it was a casual walk-in test administered at welcoming FAA facilities. The FAA is no longer interested in meeting the general flying populace, or providing much in way of service; the knowledge test is now given by private contractors, for a stiff fee, by appointment and only at computerized testing centers. Hence, it can no longer be properly called a "written" since the use of paper is passé.

Given the expense and time involved, it pays to prepare well. On the face of it, the knowledge test looks easy enough; a minimum score of 70% is passing, and there are just 60 multiple-choice questions, each with but three possible answers. But the logic in the wording often challenges the slow of wit. It takes some preparation just to understand how to take the test. Two generations ago, the written was a series of true-false questions; today, much more thorough knowledge is needed.

Once the FAA figured out that some enterprising companies were paying testees to write down a few of the questions, word-for-word, and assembling them into crib-sheet study guides, it was logical to release the actual bank of test questions, first directly and then through training purveyors, who can supply not only the correct answers but reasoning behind them. However, it has been many years since the FAA has made the database of knowledge test question public knowledge. FAA knowledge exams are now considered "closed tests," meaning no one other than the FAA has access to the actual database of questions.

Several providers do still provide a valuable database of representative questions that test on the same aeronautical knowledge, but the majority of the questions will differ slightly in the way they are asked or the answer stems provided. This makes it even more important for the student to understand the aeronautical knowledge outlined in the Airman Certification Standards and not simply try to memorize the representative questions. If the test is failed, or you just want to improve your grade, it can be retaken after additional training, but when you do you will face a different series of questions from one of the many FAA form tests in the system.

There was once no instructor's recommendation needed to take the exam. When it was giving the test itself, the FAA began to resent having the same applicants coming back time and again to waste its time, so a flight or ground instructor's endorsement, attesting to the student's readiness, began to be required. Although the pay-to-play private testing centers could probably care less, the signoff is still necessary.

The bank of test questions is continually updated, not always with questions of equal value. No longer a true evaluation of the knowledge needed to be a safe pilot, the written has become a puzzle quiz with quirks of logic in the wording, sometimes involving a backwards order of solving for an answer and a focus on reading the fine print of on-screen graphs and tables. A private pilot is now asked questions that were once reserved for the commercial pilot exam, like operation of a constant-speed propeller. And until recently, the ancient and honorable art of ADF navigation lived on, at least on the test.

PREPARATION

Unfortunately, much of the preparation for the modern knowledge exam is spent on just learning how to take the test. Students must realize that, vital as the knowledge of flight is, a good score depends on understanding how to read the sometimes arcane questions and how to sort out the wrong answers. I do not believe in memorizing mounds of material and regurgitating it on command. It's important to understand basic theory first, and then apply the testing principles to those facts.

Students vary widely in their learning style. Reading the printed page is not everyone's strong suit, even though the most trendy, tech-savvy students still rely on their reading skill to bring the words to the brain. For those who find printed prose boring, watching downloaded presentations can serve adequately, at least as a motivator to study. Even more helpful are the computer-based interactive training programs, where clicking on an icon amplifies a subject or choosing a wrong answer brings up an explanation of why the response was incorrect.

By understanding the "why" underlying the answer, a trigger is set that can go off when a wrong option is presented. A memorized response, without deeper knowledge, is an easy set-up for picking the almost-right, but always-wrong, answer.

There are marginal notes and information that are important, particularly on questions covering the weight and balance and performance graphs. It's important to look for all this information and not to jump to a conclusion. For example, the temperature aloft needed to solve a question may be presented in relation to ISA, in actual degrees or in calibrated form.

As with any timed test, it's important to move along through the easy questions first, skipping the time-consuming ones but making a careful list of them so you can return later. Each answer counts for the same percentage of the grade; don't give away a quick point or two by hanging up on a complex calculation and running out of time before completing all the questions.

EVALUATING READINESS

When determining a student's readiness, there's nothing better than a sample test, even if it's one that's been taken before. As long as a few days have elapsed since you last took it, a sample test serves to gauge the retention of knowledge and testing skill. Naturally, the more frequently a sample test was taken, the higher the score should become. Most importantly, a repeated failure of the same subject area calls for zeroing in on with some one-on-one tutoring.

If no sample test is handy, just pick out some random questions from the study booklet, pacing from subject to subject, covering up the correct answer and explanation. If the student shows equal aptitude in all the areas, they are ready for the real thing. Do not blindly sign off a student just because they profess to be ready, figuring it's their loss, not yours. You have a duty to assess their readiness. At the current inflated fee charged for administering the test, let alone the expense of travel and time involved, no one should take the test twice.

Once the sacred embossed score sheet is in hand, the job isn't done. The missed subject areas, cryptic as they are, need to be reviewed the very same day, if at all possible, while the questions are still fresh in mind. That way you can correct the false knowledge that caused the subject to be missed, rather than have the student carry it on into a flying career. You don't want the student to bring the same lack of knowledge into the oral portion of the practical test either. If the test was flunked, you'll need to go over the subjects failed in even greater detail before you sign off the student to take the test again.

Teaching the academics is not every flight instructor's strong suit. Most of us want to fly, not lecture from a manual. But teaching flying is not just about moving the controls, it's also about applying knowledge to a situation. At the very least, supplement your classroom teaching skills with multi-media sources, then do your job of evaluating the student's learning.

LESSON PLAN

OBJECTIVE

To attain a score on the actual knowledge exam of 80% or better, through taking the practice exams until achieving over-90% scores repeatedly.

TECHNIQUE

Use self-study to gain most of the knowledge and skills, with an instructor assisting with practice test questions that are missed repeatedly. Concentration will focus on areas that are obviously deficient. Emphasis will be placed on reading the test questions carefully.

DESIRED RESULT

The student will be able to read and understand test questions and pass the exam on the first attempt.

CHAPTER 13

ESCAPING WEATHER

Developing basic instrument flying skills, as we discussed in Chapter 3, is not a difficult task given the gyroscopic or full-glass instrument panels in today's airplanes. Student pilots quickly learn where to look for attitude and performance indications, all too often spending an inordinate amount of time with their eyes inside the cockpit, rather looking at the huge natural horizon outside. They continually have to be reminded that other planes are out there and to keep their head on a swivel.

However, a demonstration of instrument flying survival skills is part of the practical test, and a minimum of three hours of instruction in actual or simulated instrument flight must be logged in order to qualify for the private pilot certificate. Given the intensity of the task, this instruction should be given in 20-minute increments, perhaps a maximum of 30 minutes, rather than burning through it in hour-long sessions. Integrating instrument flying with other maneuvers makes it more applicable, practicing first with visual references, then donning the "hood" to do the same thing by instrument references.

The intent of the curriculum is not to create an instrument pilot but to make a VFR private pilot capable of surviving an inadvertent encounter with IMC (instrument meteorological conditions), and return safely to VMC by using the equipment in the panel. The three hours of instrument time are just enough to teach how to scan the instruments, learn primary references for each basic maneuver, and keep the airplane upright while using some navigation equipment.

If the CFI is current and qualified for IFR flight, a few minutes of "actual" flight in a convenient cloud layer can be a valuable teaching moment. With ATC approval, a quick climb into the cloud to perform some turns to a heading, hold a course and descend to break out under the cloud base will serve to show how difficult it is to control the aircraft without a view of the world outside. All the hood time in the world won't replace a blanket of gray around the windows, a restriction that can't be removed by peeking or raising the hood. Observing the gradual transition from VMC to IMC as one flies near a cloud base teaches a student the value of dropping down a few hundred feet to improve flight visibility.

As with any realistic, practical training, it's important to stress that we're teaching the dangers to be avoided, not encouraging hazardous flying. It has to be emphasized that flirting with bad weather can get you killed, and becoming a true instrument-rated pilot requires much more training than just the three hours in the private curriculum.

DIVIDING ATTENTION

Once the student has learned to hold the airplane in straight-and-level flight, make a standard-rate turn to a chosen heading, and find the attitude and power to make a climb or descent to a specified altitude, it's time to put these meager skills to work. Keeping the airplane under control, while using VOR or GPS to orient and track an escape route, requires juggling two tasks at once. I always stress, "this may save your life someday, so learn it well."

Electronic orientation, important as it is when one is lost in the clouds, must not take priority over keeping control of the airplane. Aviate first, then navigate, and then communicate when there is time. The student should take a second to twist the OBS (omni bearing selector) knob to center the needle on the local VOR, which is probably already tuned in, then go back to flying by the artificial horizon. Looking back at the VOR indicator will determine whether the course selected will take the aircraft "To" or "From" the station. A "From" indication works well for quickly orienting you to your position on line running out from the station, as shown by the numbers on the charted circle around the station, while the "To" display tells us we need to fly the indicated course to go to the station.

If a pilot is totally confused, they will probably want to head for the station and find their way from there, so a "To" indication should be selected; if the needle is centered with a "From" indication, they should

look for the "To" number at the bottom of the instrument and rotate the OBS knob to place it at the top. With a centered needle showing a "To" flag, it is then necessary turn the airplane's heading to match the course selected. Remember, "To" or "From" flags do not mean the airplane is actually doing that, unless the airplane is turned to a heading that's the same as the VOR course. If the needle deviates from center, usually due to wind drift, establish a corrected heading of perhaps 20 degrees in the direction of the displaced needle, and wait for results. The needle will not respond instantly; the airplane has to move across the face of the earth to get back on the selected course, so hold the new heading and watch the needle. If there's no response after a minute or so, apply twice the correction, and when the needle does return to center, take half the correction away. Remind the student to always fly the compass, not the VOR needle.

For GPS orientation, it necessary to establish a waypoint, selected either from the face of the display or from a list in the data base. This is where the student needs to practice with the GPS unit to become familiar with the required sequence of button pushes. For emergency use, most GPS units have a "nearest airport" function, which instantly calls up a listing of nearby fields, with the closest one displayed at the top. We may not want to go to the nearest one, however, if it's further into the bad weather. Select an airport that's safely in the clear, enter it as the desired waypoint, and look for the bearing to that point. Then, turn the airplane in that direction. Make the track readout on the GPS match the bearing and you'll be heading directly to the waypoint. Note the compass heading required to maintain that track, and make it your reference for flying the airplane. Again, remind the student to fly the compass, not the GPS.

REGAINING CONTROL

Another required task on the practical test is a demonstration of recovery from an unusual attitude, one induced by the instructor while the student has their eyes diverted from the instrument panel. I will ask for the student to look down at their lap or belly-button, then I'll establish the unusual attitude and tell them to look up at the panel and respond to my call, "you have the controls." They are to make a safe recovery to straight-and-level flight, showing their ability to survive a momentary loss-of-control in the clouds.

Many times, the student will have already experienced this on their own, while floundering around under the hood trying to learn basic instrument flying. By having to bring the airplane back to even keel, they have learned that the attitude indicator is a primary reference to regain control. However, I find it best to initially demonstrate just what is going to happen during a recovery-from-unusual-attitude task, first letting the student see it without the hood on, and to explain who is flying at each point during the entry and recovery. Seeing the maneuvers first with outside references takes the mystery out of what will be happening with the hood on.

In general, two scenarios are used in the practical exam, taken from real-life possibilities. One is the approach to a stall, caused by inadvertently pulling the yoke back to raise the nose higher and higher, causing the airplane to lose airspeed until a stall is imminent. When the examiner says, "you have the controls," the proper response is to look up, observe the nose-high attitude on the attitude indicator and a slow, decreasing airspeed indication, and promptly lower the nose to level flight, simultaneously adding at least cruising power to hasten the recovery. Commonly-seen errors are a slow response and over-correction.

The opposite unusual attitude situation is the entry to a spiral dive, during which the examiner gradually rolls into a bank and allows the nose to go down, until the bank diverges into an ever-steepening angle that pulls the nose further down; this "graveyard spiral" will, if not corrected, terminate in airframe failure at high speed. Upon taking over the controls, the student must immediately reduce power, level the wings, raise the nose to the horizon and keep the aircraft in a straight-and-level condition until the speed slows to normal cruise, whereupon power is returned to its usual cruising indication. The most common errors are a failure to cut power and allowing the nose to rise into a climb, rather than stopping it in level flight.

For realism, while the student is head-down I will begin with a left bank, then gradually roll into a right bank to induce spatial disorientation, and perhaps transition from a pull-up into a nose-down attitude, teaching that one's internal references must be disregarded in favor of the indications on the panel. Once the basic recoveries are learned, I will fly aggressive entries into left and right turns, climbing and descending, then give the aircraft to the now-disoriented student in level flight, observing their reaction. Sometimes they see that there's nothing to be done, and sometimes they react to their internal feelings without analyzing the situation, yanking the airplane into an unneeded recovery.

Basic instrument flying, covering the four fundamental flight maneuvers, navigation orientation, and recovery from an unusual attitude, is a life-saving response to an emergency situation. Private pilots need to know how to use the instrument indications in front of them, not just to pass the practical test but to survive an actual encounter with bad weather.

LESSON PLAN

OBJECTIVE

To refine the introductory instrument flight skills learned before solo, to meet and exceed the requirements of the practical test for the private pilot license.

TECHNIQUE

The simulated instrument flying will be broken up into short segments, integrated with visual training, to avoid fatigue. Once fundamental control and scan is learned, VOR or GPS navigation is added to assure that the student can find their way out of weather. Innovative entry to unusual attitudes, requiring the student to overcome disorientation, will be used to simulate loss-of-control in IMC.

DESIRED RESULT

Student must demonstrate precision heading, altitude and vertical speed control to at least twice the practical testing standard, and correctly respond to simulated ATC instructions. Orientation should result in a turn toward a desired electronic course and tracking toward the objective. At no time should a recovery from a simulated unusual attitude result in a stall or exceeding airspeed limits.

CHAPTER 14

DEMONIC DARKNESS

Night flying under VFR is not often given proper respect; an examination of the Federal Aviation Regulations will find very little added restriction, other than raising the minimum visual requirement in Class G airspace from one mile to three and mandating an extra 15 minutes of reserve fuel. A newly-rated private pilot, then, can quite legally load up passengers and launch off at midnight into a drizzly overcast with a ceiling of 1,000 feet and three miles of horizon-less visibility. When teaching night flight, it's important to stress the risks involved and how to manage our night-flying freedom safely.

Night training should begin with a lecture, delivered while waiting for darkness to fall. We are not interested in beginning the flight at twilight; night means dark, with no western glow to soften the impact. The FAA-mandated three hours of night training are barely sufficient to attain basic skills.

We start by discussing what night VFR weather should be; no clouds within reach of the aircraft's operating altitude limit, no ceiling that obscures the star canopy overhead, no precipitation present or forecast, and no strong surface winds to make takeoff and landing difficult. There must be enough ground lights to create at least some semblance of a visual horizon. The terrain height must be well known and altitude planning established that will avoid it. In other words, we want perfect weather and operating conditions, much more so than during daylight.

Because we are given such freedom to fly VFR at night in the United States, a privilege not enjoyed in other parts of the world, we must use this freedom responsibly to avoid adding to the accident rate. The fatal accident rate at night is much higher than what it is in the daytime, so we must not increase it further, lest it come to the attention of the authorities and we lose our ability to operate so freely. We can do this by adhering to weather and operating standards to mitigate the risk.

Here are some of the topics I will cover in the preflight briefing before beginning night training:

- Do not fly at night in any aircraft that's in less than perfect condition; a leaking door, a glitchy radio, an occasional rough mag—these may be tolerated during daylight, but not at night. You will hear strange noises and see (or not see) things at night that won't occur in daytime; you need a good, familiar, trusted airplane in the dark. Be able to locate switches and controls without illumination; know your airplane intimately before flying it at night.

- Don't fly late at night after having been up all day. Night flying is difficult enough to require your full attention and skills; fatigue can cause you to overlook a dim clue in the darkness. Unless you've had a sleep period, limit your night flying to terminate at 10 p.m.; do not try to leave at midnight and arrive at home at 2 a.m. after you've been awake since dawn.

- Watch the weather much more diligently than in daytime, realizing that many airports are unattended at night and your alternate options may be limited. Pay attention to the spread between the temperature and dew point; if they are a few degrees apart, and drawing closer, any moisture present will cause fog. Don't take off a couple of hours before sunup and plan to arrive in time for breakfast; fog may not be reported when you depart, but when the sun comes up it'll stir out of its bed and cover the landscape like a blanket. Many airports are built on inexpensive low-lying land, perfect locations for ground fog.

ON TO THE AIRCRAFT

Once the safety lecture is absorbed, it's time to prepare. Preflighting will require a flashlight unless the ramp is well-lit; there should be two flashlights on board with fresh batteries, a large and small one. Most students will pronounce the airplane ready to fly, but never think to turn on the required lights to see if they work. There's no point in starting up if the landing light wire is unhooked or a bulb is burned out.

We must avoid bright lights that take away our night vision. If you have only a large flashlight, cover most of it with your hand when reading the checklist. Clear the prop diligently before cranking; there may be unseen ramp personnel out there. The student should not be allowed to move until the position and anti-collision lights are on, and be ready to flip on the taxi light as the aircraft moves off the well-lit ramp; they should be able to find the necessary switches in the dark.

Remind students to taxi more carefully at night, and avoid trying to tune radios and input flight plan data while moving; it's far too easy to roll off the edge of the pavement in the dark. While running up, make sure the brakes are set to keep the aircraft from creeping forward; slow movement is harder to detect at night. Do not idle the engine with all lights and radios on; there may not be enough alternator output to keep the battery charged.

Have students pull into position and hold before making the first night takeoff. Point out the evenly-sloped converging of the runway lights when on the centerline, and their shoulder-height relationship to the airplane as it sits on the pavement. These cues will be needed later during the landing. The student should conduct the takeoff with a little extra airspeed at liftoff, and be cautioned to watch the attitude indicator to hold a straight-ahead climb until gaining a few hundred feet of altitude. Night flying is very close to instrument flying, particularly right after takeoff when the lights up ahead are not yet visible; it's easy to let the nose drift down from climb attitude or go into a slow turn.

We'll always climb to an altitude that will give plenty of time to explore options in an emergency, maintaining no less than 2,000 feet above the ground, even for a local flight. Point out to the student that you cannot "feel" altitude at night, so you must rely on the altimeter's indication to stay clear of terrain. The first hour is spent in learning how to stay oriented by lighted landmarks; freeways, airport beacons, town and industrial lights, and the antenna towers that can't be seen during the day, but now show up brilliantly.

Students need to consider their emergency options when flying at night. A night-time engine failure in a single-engine plane means descending into an invisible forced landing. Once they've switched tanks, turned on the boost pump, applied carb heat, and done any other restart procedure without positive results, the aircraft will be committed to an unknown arrival. The best chance for survival comes from having a lighted airport within glide range. Otherwise, one must head into the wind, look for an open area without lights and hope that it's not filled with trees. As the ground is neared, reduce impact by slowing the airplane down to minimum-sink speed, tighten the belts and harness, and prepare to ride out the crash.

TAKEOFFS AND LANDINGS

Once night orientation is mastered, it's time to go back to the airport for takeoff and landing practice. Have the student look for a beacon and runway lights in the darkness, or if flying from a big-city field, look for the dark hole in the city lights, which should signify the airport.

Have them use the altimeter to establish the proper traffic pattern height, and fly the pattern slightly wider, taking time to stabilize the approach. Make them verbally confirm the altitude when turning onto final, making sure it's no lower than 300 feet AGL, while watching carefully for "flickering" of the threshold lights, which means tree branches are sticking up into the line of sight. Have them use the glideslope lights to maintain a safe approach angle.

When introducing night landings, I prefer to teach students to land at first without the distraction of landing lights. During the flare, the runway edge lights should end up at just below shoulder height, evenly angled to place the aircraft in the center of the runway, at which point the airplane is allowed to settle slowly as speed dissipates, touching down nose high. We will use the landing lights only as taxi lights during the exit from the runway. That way, the student will never have to fear a failure of the landing lights, because they will know it's possible to land without them.

Once the feel of landing with only the runway lights has been accomplished, I introduce the landing lights as a practical aid for illumination of runway markings and tire marks, which aids control in crosswind situations and can be used to avoid unseen wildlife or other obstructions. However, it's important to stress the need to avoid fixation on the area shown by the landing light beam; we must continue to use the runway lights as the primary reference. We will never intentionally land without landing lights unless we've first checked the runway for hazards.

To gain the most benefit from night transition training, the student must first be competent in daytime flying, which is why I prefer to introduce it when nearing the checkride. Even so, I stress that they should restrict themselves from night operations until acquiring 100 hours or so of total time, at which point they should return to their CFI for additional brush-up night work. They will find the difficult task of night flying much easier once they've gained some experience after the checkride.

After proficiency in takeoffs and landings is attained, we will take a short dual cross-country flight to an airport at least 50 nautical miles away, to satisfy the 100-mile night training requirement, and make a full-stop landing there. The planning and execution of the trip will use pilotage and dead reckoning, just as in daytime, but with lighted checkpoints like towns and towers. It's important to have electronic navigation backing up the landmarks, which are easily confused in the dark. On a clear night, navigation is so simple that the main challenge is to avoid boredom, another reason to fly night VFR only in perfect weather conditions.

Night flying is addictive, because it's so peaceful and smooth to be sailing over a beautiful expanse of lights. Its risk must be mitigated by careful planning and conduct, never forgetting that people are diurnal animals not well suited to operating in darkness. If new pilots stick to good VFR nights, they'll always enjoy the experience.

LESSON PLAN

OBJECTIVE

To lay the foundation for a transition from daytime flying to night operations, through an objective analysis of the risks involved and ways to maximize safety. The goal is not to simply "check off" the requirements for the license, but to make the student capable of flying at night after the checkride as experience is gained.

TECHNIQUE

Insist on clear VFR weather conditions with light winds, using a well-prepared airplane. The training will take place during two or three different evenings. Planning the night flights is even more important than in daytime flying. Constant quizzing during the flights should make sure the student can identify landmarks and use navigation aids and compass headings to maintain orientation.

DESIRED RESULT

The student will be able to take off and land at night without assistance from the instructor, and will be able to locate airports during a cross-country flight. A navigation log will be maintained and a straight course will be flown without deviations.

CHAPTER 15
EMERGENCIES

A lot of what we teach during a flight training course is preparation for abnormal situations. If everything always went as expected, there wouldn't be much use for a pilot in the airplane. One of my favorite teaching questions is, "what is the purpose of having a pilot onboard?" Most students state the obvious: "to control the aircraft." But my desired answer is: "to make decisions." That's what we're there for, and nowhere it is more important than in an emergency.

Even in a simple push-pull fixed-gear airplane, a critical system failure can present an emergency, requiring pilot action to correct or mitigate. There isn't time to consult a detailed is-and-if flow chart. The pilot must at least begin corrective actions immediately, from memory, then proceed to any checklist available.

For example, most recently-designed airplanes have an over-voltage relay built into the electrical system, which will open to disable the alternator if excessively high voltage is detected at the voltage regulator. A red warning light should illuminate, labeled "over-voltage." Snapping off the master switch and flipping it back on will reset the relay, and if the over-voltage event was only a momentary spike the flight will proceed normally. If the relay opens again, the pilot must reduce the electrical load being drawn from the battery and expect total failure of the electrical system in a relatively short time. A lot of decisions need to be made.

Teaching such a scenario involves quizzing. Ask, "what happened and what action is needed if this light comes on?" Not, "what do you do if the over-voltage relay opens?" The answer will show if the student has read up on the electrical system, knows what to do, and understands the chain of failures that can follow if operating on battery only.

Students need to know the difference between emergencies, abnormalities and inconveniences. A total loss of engine power is a true emergency. A loss of one magneto is an abnormality. A failure in one of two communication radios is an inconvenience. Each one requires pilot action, but with different urgency and consequences.

IF THE ENGINE FAILS

From early in the student's training, we will introduce forced-landing drills. Anything mechanical can fail, and that includes the reliable old motor turning the propeller. Typically, the instructor yanks the throttle closed, announces "engine failure" and teaches (or watches) a response. The first priority is always to fly the airplane properly, trimming airspeed to the best-glide number while maintaining altitude as long as possible, avoiding a stall but maximizing time aloft. Second, the student must look around for a landing spot, and head for a likely choice immediately; it may soon be out of reach. Third, they should attempt to correct the power loss (focusing on air, fuel and ignition, the three elements of combustion), turn on carburetor heat, switch fuel tanks and activate the auxiliary fuel pump, and then adjust the mixture and mag switch.

In reality, engines seldom fail completely and suddenly. A gradual or partial power loss is more typically seen, requiring additional decision making from the pilot. Be sure to teach such partial engine failures, as well as engine-out scenarios. "What would you do if the engine is only developing 1,700 RPM?" I'll ask, and then see if the student will try to head for the airport. If they do, I'll stress that an alternate spot should be available if the altitude doesn't last long enough.

The student should bear in mind the surface wind's direction and the key position and altitude to be reached during an approach to the landing spot, while attempting to fly a normal traffic pattern if possible. Power should be restored and a go-around begun at a safe altitude, but not before the student can see the success or failure of the landing attempt.

As experience in handling practice forced landings is gained, be sure to include simulated engine-out landings at an airport, keeping the power at idle all the way to touchdown. The student needs to know a power-off approach and landing is possible, and practicing them is important, by learning when to extend flaps and how to keep sufficient energy in the aircraft to land well beyond the threshold.

Induce engine failures at all points in a flight, not just at high altitude in the practice area. I like to call for an engine failure when flying downwind during rectangular patterns, when there's only enough altitude to turn into the wind and aim for a field ahead of the nose. I also cut power just after a short-field takeoff, while the airplane is in a steep climb, requiring the student to quickly lower the nose to maintain airspeed, demonstrating that only a straight-ahead landing will be possible.

DECISION-MAKING

If a pilot is forced to make a decision to handle an emergency, perhaps an earlier decision could have been made to avoid the emergency in the first place. Deciding to make a fuel stop avoids running out of fuel, for instance. The inconvenience of stopping precludes the abnormality of running one tank empty, leaving only the remaining fuel in the other tank to get home, or, even worse, the full-on emergency of total fuel exhaustion. Even if the airplane has a "both" tank selection available, I teach switching to a single tank when half the total supply has been burned, leaving the other tank for reserve.

Deciding when to abandon the flight plan is the pilot's responsibility. Do not teach students to press on in fading daylight or deteriorating weather. Rather, show them, by your example, that every flight doesn't go as planned. Some of my most productive cross-country training came about when it was necessary to divert to an alternative route, because winds were stronger than forecast or the ceiling had dropped. The student learned to make a decision, not to stick to the plan.

Going around from a poor landing approach is another example of pilot decision-making. I enjoy watching a student make such a decision and execute a well-flown go-around, without consulting with me. That shows that my job as an instructor has been validated. I simply point out that one should try something different on the next attempt at landing, unless the reason for the go-around is no fault of the student's flying. Using less flaps or a longer final, adding crosswind correction or more altitude on base leg—these are decisions that make a pilot.

A window or door popping open on the takeoff roll or in-flight should be an inconvenience, not an emergency, but it will require a decision in order to handle the situation. It is necessary to induce such a noisy distraction during training, to make sure the student won't try to latch the open orifice during the takeoff run. The correct response is to abort the takeoff, if there's time, or fly the airplane around the pattern and land if the opening occurs later. Some doors or windows can simply be latched in flight, using proper technique, but only after establishing a safe flight condition.

Teach responses to as many system failures as the airplane has systems, introducing all the possible faults that can happen. Don't limit emergencies to the POH's section on emergency operation. No single failure should result in injury or death, as long as the pilot has been trained to make decisions.

LESSON PLAN

OBJECTIVE

To confirm knowledge of aircraft systems to aid decision-making in emergency situations, and to foster a willingness to make timely decisions throughout the student's piloting career.

TECHNIQUE

Use quizzing and innovative teaching scenarios involving emergencies at any point in training. Simulate emergencies safely and with emphasis on making correct decisions.

DESIRED RESULT

Student will demonstrate good decision-making on their own, using POH emergency procedures and common sense solutions.

CHAPTER 16

IN PURSUIT OF PERFECTION

As we near the conclusion of our training, all aspects of primary aviation have been covered. Our student has passed the knowledge exam, completed the mandated training hours in each category, and shown aptitude in all aspects of piloting. It is now time to prepare for the private pilot checkride, or practical exam. A convoluted, overly-verbose guide to the test is available from the FAA, now called the Airman Certification Standards (ACS), so all the rules of the game should be known to the parties concerned: the applicant, the instructor and the FAA-designated examiner.

Even after the extensive rewriting of the ACS, which has supposedly made the practical test more realistic by posing scenarios requiring the applicant to demonstrate the judgment required of a private pilot, there are still standards of performance to be met. In preparation for the checkride, we will hone the student's skills to at least half the minimum standards set forth by the FAA. Applicants will inevitably be nervous on test day, and a person who has only demonstrated performance barely adequate to pass the test will come up short when shaking hands and misfiring brain cells interfere with responses.

So, if the altitude tolerance for a steep turn demonstration is plus-or-minus 100 feet, we will insist on keeping altitude within 50 feet. While airspeed on landing approach may be allowed to vary by +10/-5 knots, I want to see the ASI needle within 2 knots of the target. Perfection is our goal; while we can't realistically achieve flawless performance, if we continually pursue perfection, we will get better and better. "Just good enough" should not be seen as satisfactory.

The instructor should recognize that these final periods of preparation for the flight test are probably the last times they will be able to bequeath new knowledge to this individual. The student is now able to absorb more detailed instruction than before; more progress will be made in these last 10 hours than in all the previous 40 hours. If there are any tribal secrets to be passed along, if the student is to be encouraged to attain the highest possible level of perfection, if you want to impart any last-minute words of wisdom, this is the time.

If at all possible, engage the services of another CFI to ride with the student for a "second opinion." Larger schools may already have done this, as "phase checks," but if the student has had only has one instructor for the entire training period, it is very worthwhile to gain the experience of a strange voice coming from the right seat. Having become accustomed to the terms and mannerisms expressed by only one familiar CFI, the applicant may be tripped up by another person's way of speaking or responding. And the other CFI may notice some deficiencies that have been overlooked by the recommending instructor.

My usual pace for test-prepping is to ride with the student for an hour of dual, simulating the checkride. After that, one hour of solo practice is assigned, concentrating on improving the deficiencies uncovered on the pseudo-test. Then, another hour of dual is scheduled, going over the flight test again, followed by another solo period. Each time, a concentrated session of ground school is conducted, in preparation for the oral portion of the test.

KNOW THE MATERIAL

All too often, CFIs spend too much time on flying and not enough on ground training. It's important to dig into the details of acquiring knowledge and using resources if we are to create a capable private pilot. I've always concentrated on three main areas that should carry the day for passing the checkride: knowing the aircraft's operating handbook forward and backward, being familiar with the regulations that apply to private pilots, and being able to identify and explain everything on the sectional aeronautical chart. If these three resources are familiar and understood, the oral portion of the test should be a breeze.

In addition, the subject areas that were failed on the knowledge test must be gone over to make sure the student no longer has deficiencies in those topics. It's important to understand the difference between the knowledge and practical tests. The written exam tests the applicant's ability to memorize and interpret data, choosing the best of three answers presented. The practical test, on the other hand, demonstrates that the applicant can respond extemporaneously, not necessarily with an exact word but with an answer that is substantially correct, showing that they understand and can explain the material. Using the POH or maintenance logbooks as a reference may be necessary. The important thing is to know *why* one follows a procedure or *why* a regulation says what it does, not just to regurgitate data.

BECOMING A PRIVATE PILOT

When it comes to test-prep flight training, the instructor must alternate between evaluation, playing the silent role of the examiner, and instruction, pointing out the reasons the student isn't obtaining the desired performance. Students can become accustomed to having the instructor tell them when to start a maneuver, seeking CFI approval to begin a takeoff roll or adjust power to demonstrate a stall. We have to wean them away from this reliance on the right seat; they are becoming private pilots, and should be making their own decisions by now.

I am not above challenging the student with traps that they should be able to avoid. If I reach the aircraft before them, I have the right to perform sabotage; I might remove one fuel cap, or stick a straw in a static port, or leave the fuel selector in the "off" position. If these aren't caught on the preflight inspection, a lecture will be earned. One of my favorite tricks is to turn all electrical switches to "on," so a uniform presentation will be seen as the checklist is read. Almost without fail, the student will neglect to note the wrong orientation and will attempt to start the engine with the pitot heat, dome light, landing light, etc., drawing unwanted amperes.

In flight, I play the role of the distracting passenger, attempting to induce faulty flying. I might say, "is that a car wreck over there?" Or remark, "look, there are sunbathers out on the beach today!" I may purposely input the wrong VOR frequency before I ask the student to perform an orientation, to see if they will verify the station identification, or if they will slavishly head the wrong way to go home. Alternatively, I'll cover the nav frequency display with my hand, so they won't automatically know which station I'm asking them to home on.

I like to purposely ask for a steep turn demonstration immediately after evaluating turns-around-a-point. The student is likely to check for traffic and roll into a 45-degree bank while still at traffic pattern altitude, instead of first climbing above the 1,500 feet AGL minimum required for steep turns. "You are in charge of the aircraft, as the pilot-in-command," I remind them. They should have told me, "give me a few minutes, I'll have to climb to a safe altitude before I can do that."

If the student is persistently lax in clearing airspace before initiating a maneuver, I'll lock the control yoke in neutral with my hand, so the wheel won't move until they have looked outside. Similar immobilization will be implemented if they neglect to make a verbal acknowledgement of transfer of controls.

Now is the time to emphasize smooth use of the controls, preparing for the day when the new pilot will be taking passengers aloft on their first flight. Liftoffs should come after the nose is placed in climb attitude, with the airplane floating off the ground instead being yanked suddenly into the air. Power changes should take a second to apply, rather than being implemented with a sudden burst of throttle. Turns and trim use should come smoothly, not with a rough twist. Such nuances could not have been accomplished a dozen hours ago, but now the student's experience can be called upon, enabling them to fly with greater smoothness and precision.

Before graduation, we will show the almost-pilot the real purpose their previous training. I have permission to use some challenging private strips that are *real* short fields which are also narrow and obstructed, where precise takeoff and landings are a necessity. I have a superbly-maintained turf strip we can visit, and also one that is not so well-kept, where soft-field landings and takeoffs are vital to the health of the landing gear. Wind direction at these remote strips must be obtained by observing ripples on lakes or smoke trails. Simply navigating to such a simple circle on the chart requires careful map reading.

By now, the student should be able to successfully cope with the loss of any item in the aircraft. A no-flap landing merely requires an adjustment in pattern size and approach speed, and if the directional gyro tumbles (simulated by covering it up), the magnetic compass can suffice. Most importantly, the young pilot will someday find an airspeed indicator inoperative, when a spider builds its nest in the pitot tube. I'll cover the ASI at liftoff and announce the need to land without it, using power and attitude to control the aircraft. Students need to know it can be done, so they won't panic when an ASI failure eventually happens.

These are the most valuable hours in the training curriculum, when we are working with a receptive student that's no longer a beginner, but is about to become a fellow pilot. Enjoy these last opportunities to impart life-saving tips and other things you've learned in your own career.

LESSON PLAN

OBJECTIVE

To bring the student's skills and knowledge up to those of a capable private pilot and beyond, not just to pass a checkride but to survive a lifetime in the air.

TECHNIQUE

Shift from mere instruction to observation of self-directed performance, with a shared critique afterward, so that the student becomes their own instructor. Demand more than just-passing performance; cut the ACS tolerances in half.

DESIRED RESULT

To produce a pilot who will continue self-improvement throughout their career, far beyond mere private pilot standards. They will be able to survive a lifetime of flying because of what you taught them.

CHAPTER 17

THE FINAL TEST

Now that we have fulfilled all requirements and added our personal touches to create a superior candidate for the private pilot checkride, it is time to make the appointment with an FAA Designated Pilot Examiner. While FAA Inspectors are still empowered to conduct flight tests for a rating, they generally limit their testing activity to first-time flight instructor applicants, the last bastion of official monitoring of the flight training industry.

Examiners, as private individuals, must charge a fee for their services, and considering the amount of red tape and training they must endure to keep their designation, the amount is fairly reasonable. Exorbitant gouging will cost an examiner business, as CFIs will tend to use other designees that charge a more reasonable rate.

As the recommending instructor, you will know when the student has achieved the ability to act as a private pilot. The Examiner is merely going to verify what you have already determined, as an independent cross-check of your work. If you've done your job, the checkride should be nothing more than another demonstration of what the student knows and can do. Students sometimes come to me with a request to use a particular examiner, rather than the one I habitually send my graduates to; my feeling is that it shouldn't matter who they ride with, so I have no objections. The ACS checklist and standards have to be applied identically everywhere, so the outcome is going to be the same.

The appointment should be made by the CFI, not the student. Beware of telling a student that they are doing good work and are about ready for the checkride. When they hear that, they will usually stop practicing and will no longer keep up a regular training schedule. After all, you said they're done, right? Flying regularly is important to polishing checkride skills, as is quizzing for the oral portion. Do not let weeks go by without training, just because an appointment has been made. If the flight test is cancelled due to weather or scheduling delays, be sure to keep up the pace of regular flying.

Should the instructor go with the student to the checkride appointment? It shouldn't be necessary; if the applicant has been properly prepared, there's nothing the CFI can do other than provide moral support. As a private pilot, they won't have the instructor to lean on any more, and therefore there's no reason to take the CFI along. However, I always keep my schedule open in case the weather might be too marginal for a low-time pilot, but good enough for me to help get the applicant to the appointment. Some instructors like to introduce their students to the Examiner in advance, just to show them that they don't have horns and a pitchfork.

BE PREPARED

Do not trust the student to have everything in order; check the complete file of necessary items so the checkride can be conducted smoothly. The aircraft maintenance records should be in the aircraft, so the applicant can show the Examiner that all inspections are up to date. A weight and balance calculation should be made in advance, and if the Examiner has assigned a cross-country trip for testing purposes, it should be laid out using the student's current sectional and *Chart Supplements U.S.* If the student is making the trip to the Examiner alone, don't forget to endorse the logbook for that final solo cross-country flight.

As the recommending instructor, you will use the FAA's tedious IACRA (Integrated Airman Certification and Rating Application) website to complete and sign FAA Form 8710-1. I would print out two copies after completion and have the student carry one along, keeping the other for your files. The student must have a photo ID, the student license, the medical certificate and the embossed copy of the knowledge exam test results. The student license and test results will be kept and submitted by the Examiner, if the checkride is successful.

Endorsements to be made in the student's logbook are an attestation that you have given the requisite three hours of dual in the past two months, with a certification that all areas found deficient on the knowledge exam have been satisfactorily reviewed, along with showing that the student is qualified and recommended for the test, per 14 CFR §61.107 and 61.109.

Be sure the student is prepared to pay the Examiner's fee, in whatever form the Examiner desires. Most of the time a personal check is acceptable, bearing in mind that the paperwork going to the FAA could be retained by the Examiner until the check clears.

The tension felt by the student is matched by the instructor's stress level, in most cases. Something can always go wrong, even though it's not the fault of the applicant. Therefore, we always await the outcome on pins and needles. Three results are possible: the much-desired issuance of a temporary license, qualifying the new private pilot for a lifetime of learning; a "pink slip" of denial, signifying that some additional training and testing must be done before the license can be issued; or a "discontinuance," which is used when something happens that requires the test to be halted before completion, such as bad weather, a mechanical failure or other glitch.

If the test is not fully completed for any reason, credit will be earned for the tasks that were satisfactorily completed; when the retesting is done, only the portions found unsatisfactory or not completed need to be accomplished. Needless to say, any retesting should be scheduled and finished as soon as possible, to avoid losing the fine edge of preparation that has already been honed.

Checkrides are the payoff for all the hard work that has been put in by the student and instructor. Never schedule the ride before you are absolutely certain that the individual is fully qualified, even if you think the end is in sight and a few more hours will bring the stars into alignment. You only get to finish a student once. If you are lucky, they will return for additional training after attaining the license...but don't count on it. You mustn't let an opportunity pass to add one more tidbit from your storehouse of knowledge and experience. You may never get another chance.

LESSON PLAN

OBJECTIVE

To send a qualified private pilot candidate to the Designated Pilot Examiner, verifying that all the paperwork is in order. From this point on, the student will be another licensed pilot, capable of taking unsuspecting passengers aloft. Be sure they are ready.

TECHNIQUE

Go beyond the bare essentials of the regulations, making the checkride just another evaluation like you've been giving all through the preparation phase. Inspire confidence.

DESIRED RESULT

The creation of a new private pilot, worthy of your signature on the recommendation. Your satisfaction in having done a good job means more than all the monetary pay and accolades.

INDEX

A

abnormal situations 127
academics 106
accelerated stall 20
accumulated wisdom 77
ACS checklist and standards 143
adjusting speed 13–14
advanced maneuvers 68
advisory radio calls 43
Airman Certification Standards 135
airspeed indicator inoperative 138
alternate 128
approach 49
approach control 98
approach flaps 43
approach speed 42
approach stall 19
artificial horizon 35
attitude 5
attitude indicator 35
aviate, navigate, then communicate 89

B

balloon 51
base leg 43
bee-line 90
brakes 6

C

carb heat 19
carburetor heat 15
centerline 50
challenging private strips 138
checklist 6, 42
checkpoint 80
checkride 60
Class D airports 97
clearing airspace 138
climb 4
courseline 80
crab angle 26
credit card 96
cross-country flying 78
crossed controls 53
crosswind 44
crosswind leg 45

D

dead reckoning 79
decision-making 129–130
decisions 127
departure path 45
departure stall (power-on stall) 18
descent 36
Designated Pilot Examiner 143
directional gyro 8
discontinuance 145
disorientation 90
door popping open 130
downwind leg 42
drag 14

E

E6-B 79
electronic navigation 79
embossed score sheet 106
E=MC² of aviation 5
emergency 15, 127
emergency options 121
endorsements 96, 144
engine failure 128
ETA (estimated time of arrival) 81
evaluation 137
Examiner's fee 145

F

FAA 104
fatigue 120
fear 16
final approach 42
first solo flight 59
fixation 34
fixed-pitch propeller 33
flap 14
flare 50
flare out 51
flashlight 120
flight computer 92
flight following 87
flight log 78
flight plan 96
flight planning 78
flight service 88
flight service station 82
fog 120
forced landing 15–16
fuel exhaustion 130
full-stop 53
fundamentals 3

G

glide 5
g-load 69
go-around 44
GPS 89
graveyard spiral 35
grease job 52–53
ground handling 6
ground lights 119
ground references 25
ground shyness 51
groundspeed 26

H

headwind 26
hood 34
"hover" along the centerline 44

I

IACRA 144
IMC 111
instrument flying 34, 111
instrument meteorological conditions 111
instrument-rated pilot 112

K

key position 42
knowledge test 103

L

lake 138
landing 41, 49
landing approach 42
landing light 120
level flight 4
liftoff 61
lighted landmarks 121
limitations 88
longitudinal axis 50
lost 90

M

magnetic compass 8, 33
maintenance records 144
manage risk 3
maximum lift 72
minimum-sink speed 121
minimum standards 135
mixture 15
moral support 144

N

navigation 79, 87
nearest airport 113
night flying 119
night landings 122
night takeoff 121
no-flap landing 138
nosewheel 53
NOTAMs 78
notebook 73

O

obstacle 71
omni bearing selector 112
180-degree turn 36
oral portion of the test 136
over-banking tendency 69
over-voltage relay 127

P

panic 16
perfection 59
P-factor 17
phase 77
phase checks 136
pilotage 79
pilot-in-command 67
pink slip 145
plotter 80
post-solo 62
power changes 138
power-off stall 17
power-on stall 17
practical exam 135
practice area 41, 67
pre-solo maneuvers 67
private pilot 143
private pilot candidate 146
propwash 14
pursue perfection 135

R

recommending instructor 136
rectangular course 25
re-evaluation 68
refuse to solo 61
relapse 60
risk management 15
rogue traffic 61
rudder 7
runway lights 122

S

sabotage 137
sample test 106
seat pressure 70
second opinion 136
sectional aeronautical chart 83
self-fueling 97
short-field 71
short-field landings 71
short-field takeoffs 71
signature 96
simulated engine-out landings 128
simulated instrument flying 116
sink 52
slow flight 13
smoke 138
soft-field 71, 72
solo cross-country 95
solo flight 59
solo practice 62
spin 19
stall 16
stall recovery 13
star canopy 119
starting gate 90
static port 137

steep-bank turns 68
student pilot 97
S-turns across a road 27
SUA (special use airspace) 78
suction gauge 103
supervision 67

T

tachometer 33
tail wind 26
takeoff 6, 41
takeoff-and-departure stall 17
takeoff run 50
taking charge 60
taxi light 121
temperature and dew point 120
temporary license 145
TFRs 78
too much dual 61
touch-and-go 54
touchdown target zone 72
tower-controlled 41
track readout 114
traffic pattern 41
traffic pattern practice 68
tribal secrets 136
trim 5
turn 4
turns around a point 70

U

uncontrolled 41
Unicom 61
unstick speed 72
unusual attitude 114
upwind of the field 68

V

vacuum system 103
VASI 42
verbosity 60
VMC 111
VSI 16
V_X 71
V_Y 6, 71

W

waypoint 89, 113
weather acquisition 78
weight and balance 105
wind drift 30
window 51
winds aloft 95
wingtip 70
words of wisdom 136
written 105
written exam 103